$$\frac{1}{c}\frac{\partial Y}{\partial t} = \frac{\partial \alpha}{\partial z} - \partial$$

$$\frac{1}{c}\frac{\partial \beta}{\partial t} = \frac{\partial Z}{\partial x} -$$

the universe

in zero words

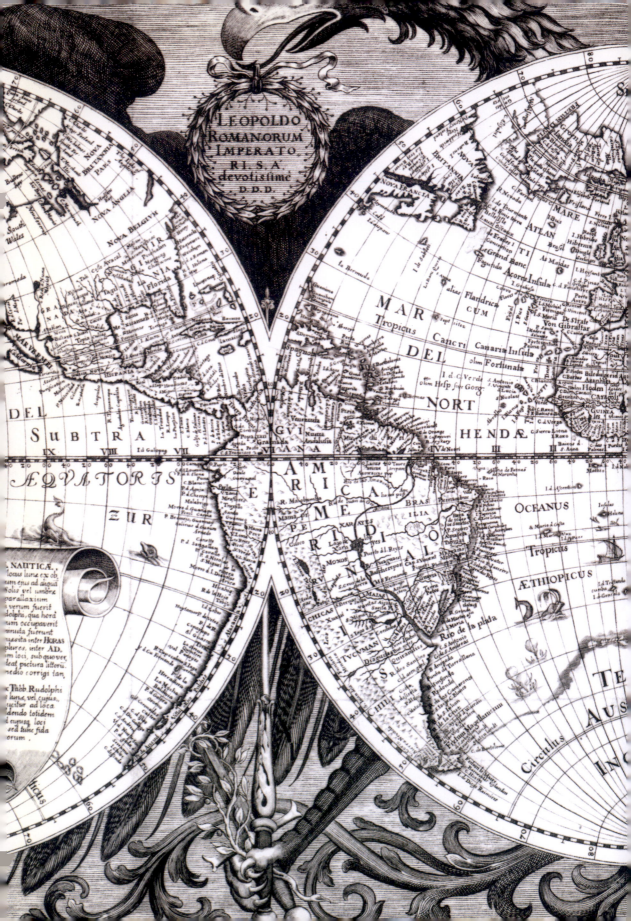

ME PRES
SAM TENEBRIS

DANA MACKENZIE

the universe
in zero words

THE STORY OF MATHEMATICS
AS TOLD THROUGH EQUATIONS

PRINCETON UNIVERSITY PRESS
PRINCETON AND OXFORD

contents

preface

In this book I hope to lift the veil of mystery and secrecy that surrounds mathematics and equations, so that those who are interested can see what lies underneath.

Firstly, let me briefly address some questions of terminology. The words "equation," "formula," and "identity" are all used in mathematics, and have slightly different shades of meaning. "Formulas" tend to be a little bit more utilitarian; you use a formula to solve an equation. "Identities" are somewhat less deep and have the connotation of something that can be proved purely by symbolic manipulation. For the purposes of this book, though, I am not going to insist on any such distinctions.

You will also frequently encounter the words "axiom," "theorem," "hypothesis," and "conjecture," in this book. An axiom is a statement that mathematicians assume as an unproven fact. They may do so because they genuinely believe it is a universal truth, or they may do it just as a convenient starting point.

A theorem is the gold standard of mathematical truth; it is a statement that has been formally deduced from a specific axiom system. It is not subject to experimental error or intellectual fashion ... except for the fact that the axiom system itself may go out of fashion. Revolutions do occur in mathematics. Usually they occur not because theorems are incorrect but because the assumptions they are based on are judged to be too restrictive, too loose, too imprecise, or not close enough to reality.

A hypothesis or conjecture (the words are synonymous) is a mathematical statement that has not been proved yet, but has substantial evidence in its favor. The evidence may come from similar but weaker theorems, empirical observations or computer experiments. Nevertheless, in mathematics a fact can never be proven by empirical evidence, plausibility, or a statistical test. This is a rule that distinguishes mathematics from the empirical sciences including physics, biology, and chemistry.

THE CHOICE OF EQUATIONS was necessarily a matter of individual taste and preference. Some equations are almost obligatory, such as Einstein's equation $E = mc^2$, probably the most famous equation of all. Other equations will be unfamiliar to all but the most savvy readers, such as the Continuum Hypothesis. Here are some of the criteria I have used to decide what makes an equation great.

1. *It is surprising*. A great equation tells us something that we did not know before. It may look like a work of alchemy, transforming one quantity into another one that at first seems completely different, yet every step can be explained and justified. The only magic is in the human mind that can discover such connections.

2. *It is concise*. A great equation has the spare aesthetic of Japanese calligraphy; it contains nothing but the essentials. It says something simple and powerful.

3. *It is consequential*. I discarded several equations that I consider to be beautiful, inspiring mathematics—but which in the end have significance only for a few cognoscenti. The equations that make the deepest impression are the ones that revolutionize mathematics, change our view of the world, or change the material possibilities of our lives.

4. *It is universal*. One of the great attractions of math is that an equation proven today will remain true forever. It is not subject to the whims of fashion, it is the same across the globe, and it cannot be censored or legislated.

Some of the equations presented here are not mathematical theorems, but physical "laws" or theories, for example, Maxwell's equations. Physical theories are generally confirmed by induction from data, or the "scientific method," rather than by deduction from a set of axioms. Unlike mathematical theorems, they are subject to empirical evidence and statistical testing, and occasionally, when more sensitive experiments come along, they are proved wanting.

The fact is that mathematics has two faces. First, it is a body of knowledge in its own right; and second, it is a language for expressing knowledge about the universe. If you look at equations merely as a means of conveying scientific information, then you are missing the way that mathematics can unbind our mental straitjackets. If you look at equations only as abstract nuggets of wisdom, then you are missing the subtle guidance nature gives us to ask the "right" questions.

LEOPOLD KRONECKER, a nineteenth-century German mathematician, once said "God created the integers; all else is the work of man". Although it is not entirely clear how literally one should take his witticism, historically he is far from alone in suggesting a divine origin for mathematics. In ancient Mesopotamia, it was a gift from Nisaba, the patron goddess of scribes. "Nisaba, the woman radiant with joy, the true woman, the scribe, the lady who knows everything, guides your fingers on the clay," wrote a scribe in the twentieth century BC. "Nisaba generously bestowed upon you the measuring rod, the surveyor's gleaming line, the yardstick, and the tablets which confer wisdom." On Babylonian mathematical tablets, the solution to a problem was never complete until the solver wrote, "Praise Nisaba!" at the end.

According to the ancient Chinese, the originator of mathematics was Fu Xi, the legendary first emperor of China. He is often depicted holding a carpenter's square. "Fu Xi created the eight trigrams in remote antiquity to communicate the virtues of the gods," wrote the third-century mathematician Liu Hui. In addition, he says, Fu Xi "invented the nine-nines algorithm to coordinate the variations in the hexagrams." The "trigrams" and "hexagrams" are the basic units of Chinese calligraphy; thus, in a loose sense, Fu Xi is being credited with the invention of writing, while, the "nine-nines algorithm" means the multiplication table. Thus, mathematics was not only divinely inspired, but was invented at the same time as written language.

We can already discern in these accounts three distinct branches of mathematics, which have continued to flow abundantly over the centuries since then. The first branch is arithmetic or algebra, the science of quantity; the second is geometry, the science of shape; and the third is applied mathematics, the science of translating mathematics into solutions to concrete problems of engineering, physics, and economics.

A fourth wellspring is not apparent in the above quotes, and that is the science of the infinite—the analysis of both infinitely large and infinitely small quantities, which are essential to understand any process of continuous motion or change. Mathematicians simply call this branch of mathematics "analysis," even though the rest of the world interprets this word to mean something quite different.

Thus, I consider the four main tributaries of mathematics to be Algebra, Geometry, Applied Mathematics, and Analysis. All four of them mingle together and cooperate in a most wonderful way, and witnessing this interaction is one of the great joys of being a mathematician. Nearly every mathematician finds himself or herself drawn more to one of these tributaries than the others, but the beauty and power of the subject undoubtedly derives from all four. For that reason, the four chapters in this book each have a theme, or "storyline" running throughout, relating to the evolution of the four branches over the ages.

introduction

the abacist versus the algorist

One afternoon in Rio de Janeiro, the Nobel Prize-winning physicist Richard Feynman was eating dinner in his favorite restaurant. It wasn't actually dinnertime yet, so the dining room was quiet … until the abacus salesman walked in. The waiters, who were presumably not interested in buying an abacus, challenged the salesman to prove that he could do arithmetic faster than their customer. Feynman agreed to the challenge.

At first, the contest wasn't even close. On the addition problems, Feynman wrote, the abacus salesman "beat me hollow." He would have the answer before Feynman even finished writing down the numbers. But then the salesman started getting cocky. He challenged Feynman to multiplication problems. Feynman still lost to the abacus, but not by as much. The salesman, not satisfied with his narrow margin of victory, challenged Feynman to harder and harder problems, and got more and more flustered. Finally he played his trump card. "*Raios cubicos!*" the salesman said. "Cube roots!"

Obviously, by this point the competition was more about pride than about selling an abacus. It's difficult to imagine why a restaurant manager would ever need to compute a cube root. But Feynman agreed, provided that the waiters, who were watching the competition and enjoying it immensely, would choose the number. The number they picked was 1729.03.

The abacist set to work with a passion, hunching over the abacus, his fingers flying too fast for the eye to follow. Meanwhile, Feynman writes, he was just sitting there. The waiters asked him what he was doing, and he tapped his head: "Thinking!" Within a few seconds, Feynman had written down five digits of the answer (12.002). After a while, the abacus salesman triumphantly announced "12!" and then a few minutes later, "12.0!" By this time Feynman had added several more digits to his answer. The waiters laughed at the salesman, who left in humiliation, beaten by the power of pure thought.

Like all good tales, Feynman's duel with the abacist has many layers of meaning. On the most superficial level, it is a story about genius; the Nobel Prize winner beating the machine. However, Feynman's intention when he told this story about himself was quite different. He was not a boastful man. In the context of his book, the point of the story was that *ordinary people*— not Nobel Prize winners, not geniuses—could do just the same thing as he did, with a little bit of number sense and mathematical knowledge. There were two secrets behind his seemingly magical feat. First, he needed to know that 1728 was a perfect cube: $12^3 = 1728$ (not common knowledge, perhaps, but it's something most physicists would be aware of, because a cubic foot is 12^3 or 1728 cubic inches.) And he needed to know a famous equation from calculus, called Taylor's formula—a very general approximation method that allows you to go from the exact equation:

$$\sqrt[3]{1728} = 12$$

to the approximate equation: $\sqrt[3]{1729.03} \approx 12.002$

Equations are the lifeblood of mathematics and science. They are the brush strokes that mathematicians use to create their art, or the secret code that they use to express their ideas about the universe. That is not to say that equations are the *only* tool that mathematicians use; words and diagrams are important, too. Nevertheless, when push comes to shove—for instance, when they have to compute the cube root of 1729.03—equations convey information with an economy and precision that words or abaci can never match.

The rest of the world, outside of science, does not speak the language of equations, and thus a vast cultural gap has emerged between those who understand them and those who do not. This book is an attempt to build a bridge across that chasm. It is intended for the reader who would like to understand mathematics on its own terms, and who would like to appreciate mathematics as an art. Surely we would not attempt to discuss the works of Rembrandt or Van Gogh without actually looking at their paintings. Why, then, should we talk about Isaac Newton or Albert Einstein without exhibiting their "paintings"? The following chapters will try to explain in words—even if words are feeble and inaccurate—what these equations mean and why they are justly treasured by those who know them.

Let's go back now to Richard Feynman and that abacus salesman, because there is more to say about them. In all likelihood, neither of them knew that they were playing out a scene that had already been enacted centuries before, when Arabic numerals first arrived in Europe.

When the new number system appeared around the beginning of the thirteenth century, many people were deeply suspicious of it. They had to learn nine new and unfamiliar symbols: 1, 2, 3, 4, 5, 6, 7, 8, and 9—or, to be more precise, they had to learn the somewhat distorted thirteenth-century versions thereof. The new symbols looked to some people like occult runes, instead of the nice solid Roman letters (I, V, X, etc.) they were accustomed to. To make things worse, they were *Arabic*—not even Christian—which made them appear even more suspicious to a deeply religious society. And finally, they included an innovation that was especially hard to grasp: the number zero, a something that meant nothing.

Nevertheless, Arabic numbers had an undeniable power. Unlike Roman numerals, which were useful for writing numbers but impractical for calculating with them, the decimal place-value system made it possible to do both. In a sense, Arabic numbers democratized mathematics. In many ancient societies, only a specially trained class of scribes could do arithmetic. With decimal notation, you did not need special training or special tools, only your brain and a pen.

The struggle between the old and new number systems went on for a very long time—well over two centuries. And, in fact, open competitions were held between abacists (people who used mechanical tools to do arithmetic) and algorists (people who used the new algorithmic methods). So Feynman and the abacus salesman were re-fighting a very old duel!

WE KNOW HOW the battle ended. Nowadays, everyone in Western society uses decimal numbers. Grade school students learn the algorithms for adding, subtracting, multiplying, and dividing. So clearly, the algorists won. But Feynman's story shows that the reasons may not be as simple as you think. On some problems, the abacists were undoubtedly faster. Remember that the abacus salesman "beat him hollow" at addition. But the decimal system provides a deeper insight into numbers than a mechanical device does. So the harder the problem, the better the algorist will perform. As science progressed during the Renaissance, mathematicians would need to perform even more sophisticated calculations than cube roots. Thus, the algorists won for two reasons: at the high end, the decimal system was more compatible with advanced mathematics; while at the low end, the decimal system empowered everyone to do arithmetic.

But before we start feeling too smug about our "superior" number system, the tale offers several cautionary lessons. First is a message that is far from obvious to most people: There are many different ways to do mathematics. The way you learned in school is only one of numerous possibilities. Especially when we study the history of mathematics, we find that other civilizations used different notations and had different styles of reasoning, and those styles often made very good sense for that society. We should not assume they are "inferior." An abacus salesman can still beat a Nobel Prize winner at addition and multiplication.

Feynman's tale exemplifies also how mathematical cultures have collided many times in the past. Often this collision of cultures has benefited both sides. For instance, the Arabs didn't invent Arabic numbers or the idea of zero—they borrowed them from India.

Finally, we should recognize that the victory of the algorists may be only temporary. In the present era, we have a new calculating device; it's called the computer. Any mathematics educator can see signs that our students' number sense, the inheritance bequeathed to us by the algorists, is eroding. Students today do not understand numbers as well as they once did. They rely on the computer's perfection, and they are unable to check its answers in case they type the numbers in wrong. We again find ourselves in a contest between two paradigms, and it is by no means certain how the battle will end. Perhaps our society will decide, as in ancient times, that the average person does not need to understand numbers and that we can entrust this knowledge to an elite caste. If so, the bridge to science and higher mathematics will become closed to many more people than it is today.

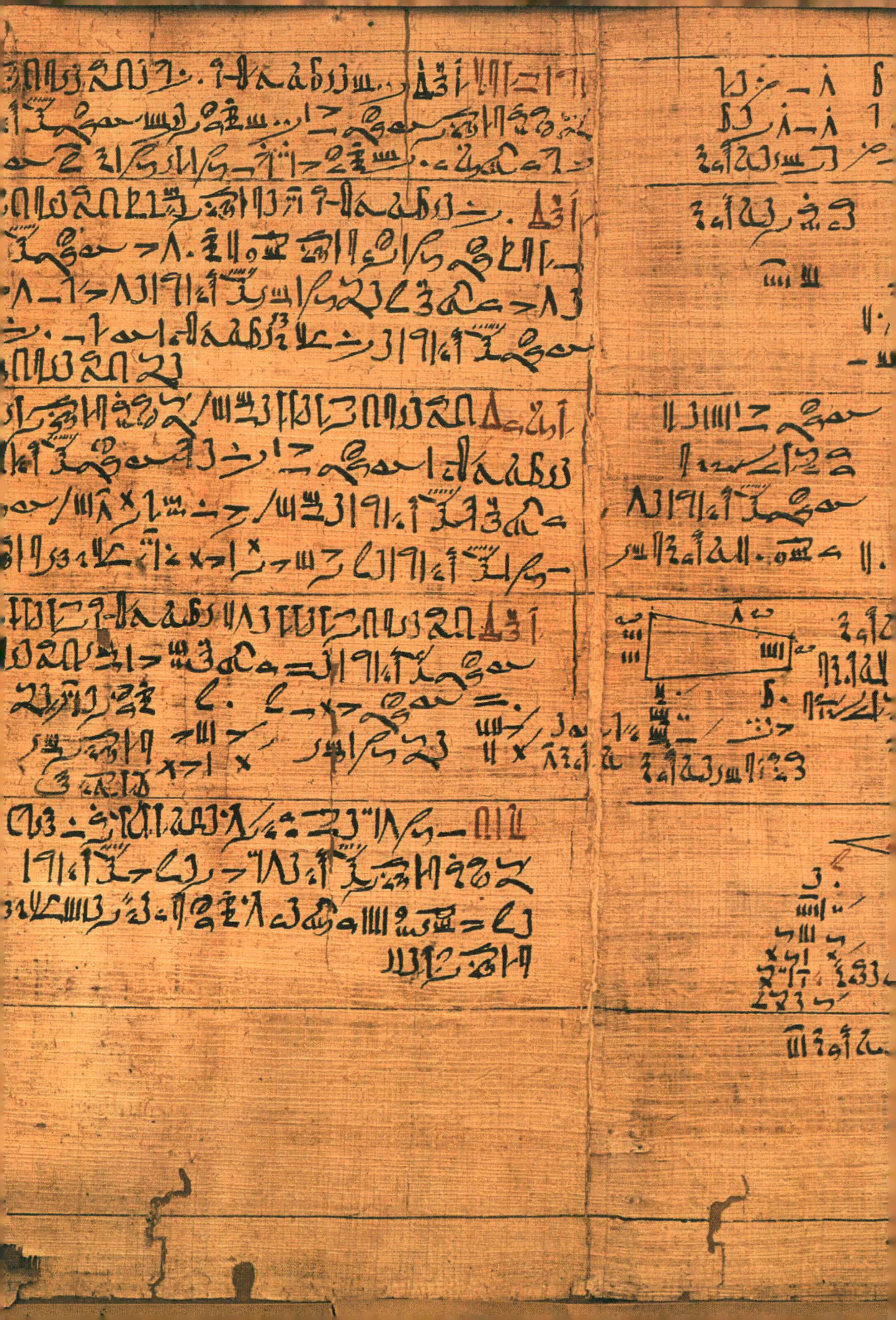

PART ONE

equations
of antiquity

In the modern world,

mathematics is an impressively unified subject. The same equations, such as $a^2 + b^2 = c^2$, will elicit recognition and understanding in any country of the world, from Europe to Asia to Africa to the Americas.

But it was not always that way. Looking back through the history of mathematics, especially in the ancient world, we see a great diversity of styles and reasons for doing mathematics. During this period, mathematics gradually evolved out of its origins in surveying, tax collecting, building, and astronomy to become a distinct subject. In Egypt and Mesopotamia, arithmetic and geometry were simply part of the scribe's general education. From the papyri and cuneiform tablets that have survived, it appears that mathematics was taught as a collection of rules, with very little in the way of explanation.

In ancient Greece, on the other hand, rote calculation took a back seat to philosophical contemplation. The Greek philosophers, starting with Pythagoras and Plato, held an exalted view of mathematics, which they saw as a science of pure reason that could penetrate behind the illusory appearance of the physical world. In Euclid's *Elements*, all of geometry is deduced from a very short list of (supposedly) self-evident facts, or axioms. This style of deductive reasoning gave birth to modern mathematics and even extended its influence to other human endeavors. (Recall the opening words of the American Declaration of

Independence: "We hold these truths to be self-evident ..."; the author, Thomas Jefferson, was laying out in true Euclidean fashion the axioms on which a new society would be based.)

In India, mathematics or *ganita* (calculation) was subservient to astronomy for many centuries, and only came into its own around the ninth and tenth centuries AD. Nevertheless, several important discoveries originated in India—foremost among them the decimal number system we use today. In China, the fortune of mathematics or *suan shu* (number art) waxed and waned over the centuries. In the Tang dynasty (618–907 AD) it was a prestigious subject that all scholars had to study; on the other hand, in the Ming dynasty (1368–1644) it was categorized as *xiaoxue* (lesser learning)! The change in attitude may have something to do with why Chinese mathematics—which was previously superior to contemporary European mathematics—stagnated after the 1300s, precisely the period when Western mathematics began to take off.

Finally, the Islamic world occupied a unique position in mathematics history, as the inheritor of two distinct traditions (Greek and Indian) and the transmitter of those traditions—augmented by the new discoveries of Islamic mathematicians—to western Europe. Strangely, it was only in western Europe that the decisive transition to modern mathematics occurred ... but that is a subject for later.

1
why we believe in arithmetic
the world's simplest equation

One plus one equals two: perhaps the most elementary formula of all. Simple, timeless, indisputable … But who wrote it down first? Where did this, and the other equations of arithmetic, come from? And how do we know they are true? The answers are not quite obvious.

One of the surprises of ancient mathematics is that there is not much evidence of the discussion of addition. Babylonian clay tablets and Egyptian papyri have been found that are filled with multiplication and division tables, but no addition tables and no "1 + 1 = 2." Apparently, addition was too obvious to require explanation, while multiplication and division were not. One reason may be the simpler notation systems that many cultures used. In Egypt, for instance, a number like 324 was written with three "hundred" symbols, two "ten" symbols, and four "one" symbols. To add two numbers, you concatenated all their symbols, replacing ten "ones" by a "ten" when necessary, and so on. It was very much like collecting change and replacing the smaller denominations now and then with larger bills. No one needed to memorize that 1 + 1 = 2, because the sum of | and | was obviously ||.

In ancient China, arithmetic computations were performed on a "counting board," a sort of precursor of the abacus in which rods were used to count

$$1 + 1 = 2$$

A simple interpretation is this: On the number line, 2 is the number that is one step to the right of 1. However, logicians since the early 1900s have preferred to define the natural numbers in terms of set theory. Then the formula states (roughly) that the disjoint union of any two sets with one element is a set with two elements.

ones, tens, hundreds, and so on. Again, addition was a straightforward matter of putting the appropriate number of rods next to each other and carrying over to the next column when necessary. No memorization was required. However, the multiplication table (the "nine-nines algorithm") was a different story. It was an important tool, because multiplying $8 \times 9 = 72$ was faster than adding 8 to itself nine times.

Another exceedingly important notational difference is that not a single ancient culture—Babylonian, Egyptian, Chinese, or any other—possessed a concept of "equation" exactly like our modern concept. Mathematical ideas were written as complete sentences, in ordinary words, or sometimes as procedures. Thus it is hazardous to say that one culture "knew" a certain equation or another did not. Modern-style equations emerged over a period of more than a thousand years. Around 250 AD, Diophantus of Alexandria began to employ one-letter abbreviations, or what mathematical historians call "syncopated" notation, to replace frequent words such as "sum," "product," and so on. The idea of using letters such as x and y to denote unknown quantities emerged much later in Europe, around the late 1500s. And the one ingredient found in virtually every equation today—an "equals" sign—did not make its first appearance until 1557. In a book called *The Whetstone of*

Wytte, by Robert Recorde, the author eloquently explains: "And to avoide the tediouse repetition of these woordes: is equal to: I will sette as I doe often in woorke use, a paire of paralleles, or Gemowe lines of one lengthe, thus: ══, because noe 2 thynges can be moare equalle." (The archaic word "Gemowe" meant "twin." Note that Recorde's equals sign was much longer than ours.)

So, even though mathematicians had implicitly known for millennia that 1 + 1 = 2, the actual equation was probably not written down in modern notation until sometime in the sixteenth century. And it wasn't until the nineteenth century that mathematicians questioned our grounds for believing this equation.

THROUGHOUT THE 1800s, mathematicians began to realize that their predecessors had relied too often on hidden assumptions that were not always easy to justify (and were sometimes false). The first chink in the armor of ancient mathematics was the discovery, in the early 1800s, of non-Euclidean geometries (discussed in more detail in a later chapter). If even the great Euclid was guilty of making assumptions that were not incontrovertible, then what part of mathematics could be considered safe?

In the late 1800s, mathematicians of a more philosophical bent, such as Leopold Kronecker, Giuseppe Peano, David Hilbert, and Bertrand Russell, began to scrutinize the foundations of mathematics very seriously. What can we really claim to know for certain, they wondered. Can we find a basic set of postulates for mathematics that can be proven to be self-consistent?

Opposite The key to arithmetic: an Arabic manuscript, by Jamshid al-Kashi, 1390–1450.

Kronecker, a German mathematician, held the opinion that the natural numbers 1, 2, 3, … were God-given. Therefore the laws of arithmetic, such as the equation 1 + 1 = 2, are implicitly reliable. But most logicians disagreed, and saw the integers as a less fundamental concept than sets. What does the statement "one plus one equals two" really mean? Fundamentally, it means that when a set or collection consisting of one object is combined with a different set consisting of one object, the resulting set always has two objects. But to make sense of this, we need to answer a whole new round of questions, such as what we mean by a set, what we know about them and why.

بسم الله الرحمن الرحيم و به نستعين

الحمد لله الذي توحد بابداع الاحاد ٭ و تفرد بتاليف صنوف الاعداد ٭ و الصلوة على خير خلقه الشافع الشافعين يوم التناد ٭ و اله و اولاده الهادين سبيل النجاة و الرشاد ٭ اما بعد فان اوج خلق الله تعالى الى عزاره جمشيد بن مسعود بن محمود الطبيب الكاشي الملقب بغياث الدين احين الله احواله يقول لما مارست الاعمال الحسابية ٭ و القوانين الهندسية ٭ حتى بلغت الى حاق نهايتها ٭ و بالغت في دقايقها ٭ و كشفت عن غوامضها و معضلاتها ٭ و حللت مشكلاتها ٭ و استنبطت كثيرا من القوانين و الضوابط فيها ٭ و استخرجت ما صعب استخراجه على كثير من مباشريها ٭ كما استانفت استخراج جميع جداول الزيج الايلخاني ٭ باوفق علو وضعت الزيج الملخص بالخاقاني ٭ في تكميل الزيج الايلخاني ٭ و جمعت فيه جميع ما استنبط من اعمال المنجمين ٭ حالا بان في زيج اخرجه البراهين الهندسية ٭ و وضعت ايضا زيج التسهيلات و جداول الشمس ٭ و صنفت رسائل اخرى مثل الرسالة المسماة بالسلم السماء في حل اشكال وقع للمتقدمين في الابعاد و الاجرام و الرسالة المحيطية في نسبة القطر الى المحيط ٭ و رسالة الوتر و الجيب

In 1910, the mathematician Alfred North Whitehead and the philosopher Bertrand Russell published a massive and dense three-volume work called *Principia Mathematica* that was most likely the apotheosis of the attempts to recast arithmetic as a branch of set theory. You would not want to give this book to an eight-year-old to explain why 1 + 1 = 2. After 362 pages of the first volume, Whitehead and Russell finally get to a proposition from which, they say, "it will follow, when arithmetical addition has been defined, that 1 + 1 = 2." Note that they haven't actually explained yet what addition is. They don't get around to that until volume two. The actual theorem "1 + 1 = 2" does not appear until page 86 of the second book. With understated humor, they note, "The above proposition is occasionally useful."

It is not the intention here to make fun of Whitehead and Russell, because they were among the first people to grapple with the surprising difficulty of set theory. Russell discovered, for instance, that certain operations with

sets are not permissible; for example, it is impossible to define a "set of all sets" because this concept leads to a contradiction. That is the one thing that is never allowed in mathematics: a statement can never be both true and false.

But this leads to another question. Russell and Whitehead took care to avoid the paradox of the "set of all sets," but can we be absolutely sure that their axioms will not lead us to some other contradiction, yet to be discovered? That question was answered in surprising fashion in 1931, when the German logician Kurt Gödel, making direct reference to Whitehead and Russell, published a paper called "On formally undecidable propositions of *Principia Mathematica* and related systems." Gödel proved that any rules for set theory that were strong enough to derive the rules of arithmetic could never be proven consistent. In other words, it remains possible that someone, someday, will produce an absolutely valid proof that 1 + 1 = 3. Not only that, it will forever remain possible; there will never be an absolute guarantee that the arithmetic we use is consistent, as long as we base our arithmetic on set theory.

MATHEMATICIANS DO NOT actually lose sleep over the possibility that arithmetic is inconsistent. One reason is probably that most mathematicians have a strong sense that numbers, as well as the numerous other mathematical constructs we work with, have an objective reality that transcends our human minds. If so, then it is inconceivable that contradictory statements about them could be proved, such as 1 + 1 = 2 and 1 + 1 = 3. Logicians call this the "Platonist" viewpoint.

"The typical working mathematician is a Platonist on weekdays and a formalist on Sundays," wrote Philip Davis and Reuben Hersh in their 1981 book, *The Mathematical Experience*. In other words, when we are pinned down we have to admit we cannot be sure that mathematics is free from contradiction. But we do not let that stop us from going about our business.

Another point to add might be that scientists who are not mathematicians are Platonists every day of the week. It would never even occur to them to doubt that 1 + 1 = 2. And they may have the right of it. The best argument for the consistency of arithmetic is that humans have been doing it for 5000 years and we have not found a contradiction yet. The best argument for its objectivity and universality is the fact that arithmetic has crossed cultures and eras more successfully than any other language, religion, or belief system. Indeed, scientists searching for extraterrestrial life often assume that the first messages we would be able to decode from alien worlds would be mathematical—because mathematics is the most universal language there is.

We know that 1 + 1 = 2 (because it can be proved from generally accepted principles of set theory, or else because we are Platonists). But we don't *know* that we know it (because we can't prove that set theory is consistent). That may be the best answer we will ever be able to give to the eight-year-old who asks why.

2
resisting a new concept
the discovery of zero

Entire books have been written about the concept of the number zero. This number was a latecomer to arithmetic, perhaps because it is difficult to visualize zero cubits or zero sheep. Even today, if you pick up a children's counting book, you will probably not find a page devoted to zero.

The number zero has two different interpretations, one of them a good deal more sophisticated than the other. First, in numbers like 2009 or 90,210, zero is used as a symbol to denote an empty place. That is the function of the zeros. Without the numeral zero, we would not be able to tell those numbers apart from 29 or 921. In a place-value number system, the meaning of "2" depends on where it is; in the number 29 it denotes two tens, but in the number 2009 it denotes two thousands.

Of course cultures that did not use a place-value system, such as ancient Egypt or Rome, did not have this problem and did not need a symbol for an empty place. The Roman numeral MMIX (2009) is easy to distinguish from XXIX (29). Thus it is not surprising that the notion of zero did not arise in those societies. The Babylonians, however, did use a place-value number system, and yet for many centuries it did not occur to them to employ a mark to denote an empty place. Apparently the ambiguity between 2009 and 29 did not trouble them—perhaps because it is usually apparent from context

$$1 - 1 = 0$$

On the number line, zero is the number
that is one step to the left of 1.

which number is intended. Even today the same thing is true. If someone is
telling you what year it is, you expect a number like 2009; if they are telling
you how old they are, 29 is more reasonable.[*]

Only around 400 BC, near the end of Babylonia's independent existence
and some 1500 years after the cuneiform number system had first come into
use, did scribes start to use two vertical wedges (∧∧) to denote an empty
place. This was the first appearance in history of a symbol that meant zero,
but it is clear that the Babylonians thought of it only as a placeholder and
not as a number itself.

THE SECOND, more subtle, concept of zero as an actual entity (as
implied by the equation $1 - 1 = 0$) arose in India. It appears for the first time
in 628 AD, in a book called *Corrected Treatise of Brahma*, by Brahmagupta.

---

[*] The Babylonians actually used a number system based on powers of 60, rather than powers of 10.
This does not alter the basic problem of ambiguity. For example, a Babylonian scribe would not be able
to distinguish 1501 (i.e., $25 \times 60 + 1$) from 90,001 (i.e., $25 \times 60^2 + 1$). Both numbers would be written as
25, 1.

As is the case for many ancient mathematicians, little information is available about Brahmagupta's life. He was born in 598 in north central India, and was a member of a mathematical school (in the sense of a loosely knit community of scholars) in Ujjain. He lived not long after the end of the Gupta dynasty (ca. 320–550), a period of prosperity that is often considered a golden age of Indian culture, when much classical Sanskrit literature was written and when astronomers developed very accurate predictions of eclipses and planetary motions.

One thing that stands out clearly in Brahmagupta's work is his derisory attitude toward his rivals. The very title, *Corrected Treatise of Brahma*, is an implicit criticism of an earlier astronomical work. Brahmagupta makes comments such as this about his predecessors: "One is not a master through the treatises of Aryabhata, Visnucandra, etc., even when [they are] known [by heart]. But one who knows the calculations of Brahma [attains] mastery." [†]

Arrogant though he may have been, Brahmagupta clearly understood the nature of zero. He wrote, "[The sum] of two positives is positive, of two negatives negative; of a positive and a negative [the sum] is their difference; if they are equal it is zero." Thus, zero is obtained by adding a positive number to a negative of equal magnitude; for example, $1 + (-1)$. This is what is meant by the modern notation $1 - 1$. Further, Brahmagupta wrote that adding zero does not change the sign of a number, that $0 + 0 = 0$, and that any number times zero gives zero. However, he is not sure about division by zero. Rather tautologically, he writes, "A negative or a positive divided by zero has that as its divisor," and he states incorrectly that "a zero divided by a zero is zero." Modern mathematicians would say that any division by zero is undefined.

It is noteworthy that zero goes hand in hand in Brahmagupta's work with negative numbers. Indeed, the resistance to zero may be explained by the even greater difficulty of visualizing negative cubits or negative sheep. For centuries after Brahmagupta, mathematicians continued to avoid negative numbers in their formulas. For example, the solution of quadratic equations and cubic equations was made unduly complicated by mathematicians'

[†] Brackets inserted by mathematical historian Kim Plofker, who translated the work.

avoidance of negatives. They perceived the need for several different methods of solution, which we now condense into one formula.

FOR MODERN MATHEMATICS it is difficult to overstate the importance of zero. It is what mathematicians call an identity element, because when added to any number it does not change that number. Identity elements are as important to mathematics as synonyms are to literature. No one would question why we need both of the words "happy" and "delighted." They allow us to say essentially the same thing in different ways, possibly revealing slightly different nuances. The availability of zero gives mathematicians the same flexibility. An expression x can be written as $x + 0$, and from there it can be rewritten as $x + 1 - 1$ or in many other ways, depending on the requirements of the problem.

In the nineteenth and twentieth centuries, mathematicians discovered many useful algebraic structures besides integers and real numbers, and useful operations besides ordinary addition and multiplication. For example, computers use modular arithmetic; cryptologists use multiplication on elliptic curves; and quantum physicists add and multiply vectors in Hilbert space. All of these operations are variations on the fundamental notions of "plus" and "times," but they are sometimes very far removed from the addition and multiplication we learn in school. The one thing that they all have in common is an identity element. Thus, Brahmagupta's contribution to mathematics—the idea of the number zero—is alive and well, even though he might have some trouble recognizing it.

3

the square of the hypotenuse
the pythagorean theorem

For many mathematics students, the first name they encounter in a mathematics course is the name of Pythagoras. The Pythagorean Theorem states that in any right triangle (that is, a triangle containing one right angle) the sum of the squares of the two shorter sides (a and b) equals the square of the longest side (c), called the hypotenuse. The formula, memorized by centuries of students, is $a^2 + b^2 = c^2$.

The Pythagorean Theorem is famous enough that it has appeared often in popular culture. In the movie *The Wizard of Oz*, the scarecrow memorably botches the theorem after the wizard awards him a diploma, exclaiming, "The sum of the square roots of any two sides of an isosceles triangle is equal to the square root of the remaining side. Oh, joy, oh, rapture. I've got a brain!" In Gilbert and Sullivan's musical *The Pirates of Penzance*, the Major General demonstrates a somewhat better command of mathematics when he sings, "I'm very well acquainted, too, with matters mathematical … With many cheerful facts about the square of the hypotenuse."

Who was Pythagoras, and what did he have to do with the theorem that bears his name? As it turns out, the answer is rather complicated. In all likelihood, Pythagoras neither discovered nor proved "his" theorem. It is long past time for the theorem to be given a more accurate name.

$$a^2 + b^2 = c^2$$

The letters *a* and *b* represent the legs (short sides) of a right triangle, and *c* represents the hypotenuse (long side).

Pythagoras was born around 569 BC on the island of Samos, off the coast of Ionia (modern-day Turkey). According to legend, he spent years absorbing the wisdom of the ancients in Egypt, and perhaps Babylonia, maybe even India for that matter (when you are a legend, all things are possible). After returning home, he then emigrated permanently to the Italian city-state of Croton. There he founded a secret society known as the Pythagoreans, which for a period dominated the cultural and civic life of Croton.

As cults go, the Pythagoreans were only mildly eccentric. Pythagoras preached the virtues of temperance, reverence for one's elders, and education. He advocated monogamy, which must have come as a shock to a society whose gods were serial adulterers. He forbade the consumption of animal flesh, because an animal might have the soul of a friend or an ancestor. In addition, he prohibited the eating of beans, possibly because human souls could migrate to these plants as well.

All in all, Pythagoras could be seen as a rather typical charismatic leader of a mystery cult. But what distinguishes the Pythagoreans, at least to a historian of science, is their alleged role as the founders of the Greek traditions of mathematics and philosophy. Pythagoras believed that everything in the world was governed by numbers.

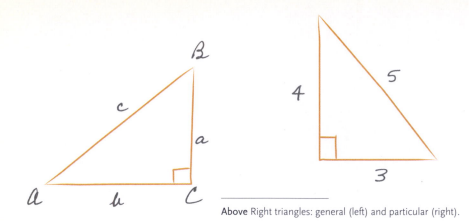

Above Right triangles: general (left) and particular (right).

Pythagorean philosophy included a great deal of numerology, or number mysticism, some of which appears laughable to modern eyes. For example, odd numbers were thought to be masculine and even numbers feminine. But their fascination with numbers did lead the Pythagoreans to some concepts that remain quite relevant to the modern subject of number theory. For instance, they discovered what they called "perfect numbers" (numbers that are equal to the sum of all their proper divisors). The first two perfect numbers are 6 (which equals 1 + 2 + 3) and 28 (which equals 1 + 2 + 4 + 7 + 14). At the time of writing, 47 perfect numbers are known, and as computers get faster and faster, they turn up new ones every few years.

AN EVEN MORE FRUITFUL concept was that of prime numbers: that is, numbers that have only themselves and 1 as divisors. The first few primes are 2, 3, 5, 7, and 11. Numbers that are not prime are called composite (for example, 6 is the product of two primes, 2 × 3).

Without prime numbers, number theory would be a relatively barren subject. With them, it is endlessly fascinating. The ancient Greeks proved that the primes never end; however, the details of how they are interspersed among the integers remain very mysterious. Primes can be used both to solve whole-number equations and to show they are unsolvable; Fermat's Last Theorem is but one example, as will be discussed in a later chapter. And finally, primes are essential for modern-day cryptography, much of which is based on the idea that it is hard to find the prime factors of a very large composite number—say, one with a few hundred digits.

Among the other mathematical discoveries attributed to the Pythagoreans were the Pythagorean theorem, and perhaps more importantly, a *proof* of the Pythagorean theorem; the principle that musical chords are formed by vibrations whose frequencies form simple whole-number ratios (for instance, an octave is created by the ratio 2:1, a fifth is created by the ratio 3:2, and a fourth is created by the ratio 4:3); the belief that the motions of the planets were governed by similar integer ratios —according to legend, Pythagoras could actually hear the harmonies that the planets produced, known as "the music of the spheres"; and finally the existence of irrational numbers.

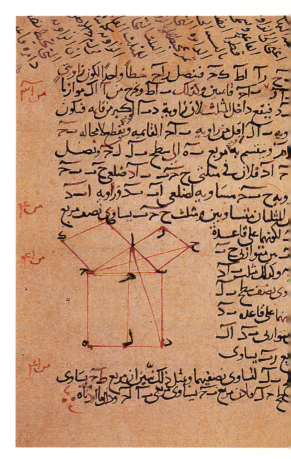

Do these claims hold water?

First, one of the few things that can be stated with absolute certainty about Pythagoras is that he did *not* discover the Pythagorean theorem. A famous Babylonian tablet known as Plimpton

Above right Pythagorean Theorem described in an early Arab book.

322, dating to roughly 1800 BC, contains a list of what we would now call Pythagorean triples: sets of integers (such as 3–4–5 or 5–12–13) that form the sides of a right triangle. (As you can readily check, $3^2 + 4^2 = 5^2$ and $5^2 + 12^2 = 13^2$.) As Pythagoras supposedly studied in Babylonia, one may surmise that he learned the Pythagorean formula there.

It would be much more interesting if Pythagoras actually discovered a proof of the Pythagorean theorem, in other words a demonstration from elementary principles that the formula $a^2 + b^2 = c^2$ holds for *all* right triangles. The Babylonians and Egyptians were apparently not interested in such mathematical deductions; their extant texts are long on procedures and short on explanations. (An explanation, for the Babylonians, is "Behold, it is done," followed by "Praise Nisaba.")

In the two centuries that followed Pythagoras, ancient Greece did develop a rich tradition of deductive mathematics, unprecedented in world mathematics. It culminated in Euclid's *Elements*, written around 300 BC. The first book of the *Elements* includes a careful proof of the Pythagorean theorem. It would be amazing, and wonderful, if the proof could be traced to the very beginning of ancient Greek mathematics.

ONCE YOU HAVE ACCEPTED, or proved, that the equation $a^2 + b^2 = c^2$ holds for every right triangle (and not just for convenient ones like 3–4–5 or 5–12–13), you run straight into a conundrum. The simplest right triangle of all is the one created by cutting a square in two parts along the diagonal. This triangle has two legs of equal length, which can be assumed to be 1 unit and 1 unit. Then, according to the Pythagorean theorem, the length of the hypotenuse, c units, obeys the equation $c^2 = 2$.

But according to Pythagorean dogma, everything in the universe is supposed to be governed by whole numbers. So this mysterious length c should be expressible as a ratio of integers, say p/q. It's easy to find some "close calls." For example $7/5$ is just a little too small, because $(7/5)^2 = 49/25 = 1.96$, and $17/12$ is just a little too big, because $(17/12)^2 = 289/144 \approx 2.007$. Thus you can say that length c is between $7/5$ and $17/12$ … But try as you might, you won't be able to find whole numbers p and q such that $(p/q)^2 = 2$.

You may wonder how I can be so sure. Let's assume that you *could* find a ratio p/q whose square is 2. Then $p^2 = 2q^2$, so p^2 is an even number. Thus p is even, and so there is some whole number x such that $p = 2x$. Thus $4x^2 = (2x)^2 = p^2 = 2q^2$, so $q^2 = 2x^2$. Thus q is also an even number, and expressible as $q = 2y$. But then $p/q = x/y$, which means we have found *smaller* numbers x and y with the same ratio. But then we could apply the same argument to x and y, getting a ratio with even smaller numbers. And so the process would never end—we could never reduce the fraction to its lowest terms! This is an absurdity, and therefore the assumption that $(p/q)^2 = 2$ must have been fallacious. This kind of proof is called a *reductio ad absurdum*, or proof by contradiction, and will be discussed further in Chapter 5.

Nowadays, we have other ways of expressing the length of the hypotenuse, c. A standard 10-digit pocket calculator says that the length is 1.414213562. But the Pythagoreans did not have a decimal notation system—decimals would not come to Europe for another 1500 years! So they did not have the option of writing the answer this way. And anyway, 1.414213562 is still not the exact length. The square of this number is 1.999999998, not 2.

A second way around this conundrum is to write $c = \sqrt{2}$. This is what is taught in school. The answer looks comfortingly exact ... but it is also vacuous. It says that the number whose square is 2 is ... the square root of 2. It tells us nothing that we didn't already know!

In any event, the Pythagoreans lacked the $\sqrt{}$ notation, and surely would not have been satisfied by such a self-referential answer. So they literally had no way of writing down the length of the diagonal of a square. It was *alogos*, a concept that could not be expressed in words. Today we would say that it is an irrational number. (A rational number is one that can be written as a ratio of whole numbers, such as $^7/_5$.) If the term "irrational" sounds a bit pejorative, that is no accident. For the Pythagoreans, if a number could not be expressed in words, it *should* not be expressed in words. The existence of *alogos* numbers should be kept as a mystery only for the deepest initiates. Legend has it that the first person who revealed the secret (possibly a Pythagorean named Hippasus) was drowned at sea as punishment.

This is a marvelous tale, but it is probably not true. It is unlikely that Pythagoras could have proved that the number $\sqrt{2}$ is *alogos*. The "proof by contradiction" technique, which lies at the heart of the argument, was introduced two generations after Pythagoras, by Zeno of Elea (a student of Parmenides, who is sometimes described as a Pythagorean but was not actually a member of the society).

Nowadays, historians place much less stock than they once did in the alleged accomplishments of Pythagoras and the Pythagoreans. There simply is no documentary evidence of it, while there is a great deal of evidence about the accomplishments of ancient Greek mathematicians who were not in the brotherhood. For instance, Theodorus of Cyrene, who also was not a Pythagorean, proved around 400 BC that the numbers we would call $\sqrt{3}$, $\sqrt{5}$, and so on up to $\sqrt{17}$ were also irrational. (The fact that he began with $\sqrt{3}$ suggests that the irrationality of $\sqrt{2}$ was already accepted by this time.)

From the viewpoint of modern historians, it makes more sense to study the many *documented* advances of ancient Greek mathematicians who were not Pythagoreans, than to perpetuate unsubstantiated legends about the Pythagoreans. One modern historian, M.F. Burnyeat, argues that the legend of Pythagoras was deliberately fabricated by Plato's successors in order to portray Plato as the heir to an ancient tradition.

There is one more good reason not to canonize Pythagoras. Science progresses much more rapidly when it is communicated openly than when it is shrouded in secrecy. As long as mathematics was hidden behind the veil of secrecy, it was impossible to separate genuine mathematics from bogus numerology. Once mathematics came out from behind the Pythagorean veil, the way was opened to new discoveries, such as those of Theodorus (and Eudoxus, Eratosthenes, Euclid, Archimedes, …). If we are celebrating ancient Greek mathematics, we should give most of the credit not to the secret cult of Pythagoras, but to the spirit of open inquiry that followed after its dispersal.

It seems a shame, also, to attribute the Pythagorean theorem to one person, when it is one of the great universal theorems of mathematics. The Pythagorean formula was discovered independently, it seems, by nearly every ancient culture. In some sense, it seems to be an inevitable discovery for any mathematically inclined civilization. If "God created the integers," as Kronecker said, perhaps he created the Pythagorean theorem too.

IN CHINA, for instance, the Pythagorean formula was known as the *gou-gu* rule. In Chinese terminology, *gou* (leg) was the shorter side of a right triangle, and *gu* (thigh) was the longer side. (Compare this to the Western terminology where both sides are simply called legs.) The hypotenuse was called *xian*, or "lute string," which may allude to the origins of the theorem in measuring distances with a stretched rope.

The *gou-gu* rule appears in the anonymously written *Jiu Zhang Suan Shu*, or *Nine Chapters on the Art of Mathematics*, a seminal work for Chinese mathematics that was as influential as Euclid's *Elements* in the West. The

age of the *Nine Chapters* is unknown. Liu Hui, a commentator in the third century AD, strongly implies in his preface that the *Nine Chapters* existed before the Chinese emperor Qin Shi Huang ordered all books to be burned, in 213 BC. After Qin's death, the *Nine Chapters* had to be reconstructed from memory. It is easy to imagine how imperfect this process was. Thus, as the work was passed down from generation to generation, a long tradition arose of improvements and comments on the original text.

Liu Hui's annotation of the *Nine Chapters* was one of the best, and contains much material of his own. Liu, a self-taught mathematician, may perhaps be considered the first Chinese math geek; he studied the subject because he cared about it and not because it would advance his career in court. In his annotations we find many explanations of why the statements in the *Nine Chapters* are true, and in particular we find the first documented proof of the Pythagorean theorem outside of ancient Greece.

Liu's argument is reminiscent of the equally ancient Chinese puzzle of tangrams, in which a small set of simple pieces is rearranged to produce a fantastic variety of figures. He starts with two squares made from the sides of the *gou* and the *gu*, and then dissects them and rearranges the pieces so

Below The *gou-gu* theorem. Liu Hui's proof shows that the squares on the *gu* and the *gou* (ABED and BCGF) can be cut into smaller pieces and rearranged to form the square on the *xian* (ACJH).

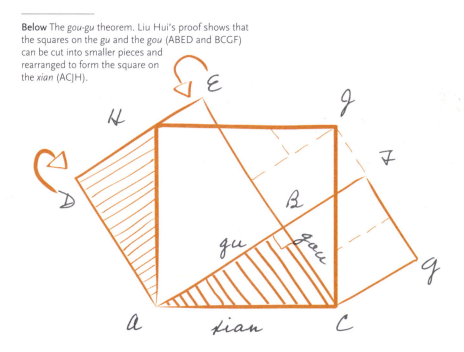

that they form a square on the *xian*. This kind of proof, called the "out-in" method, was used repeatedly by Liu and by other Chinese mathematicians. It is a much simpler proof to understand than the one in Euclid's *Elements*.

IN ONE RESPECT, the history of the *gou-gu* theorem in China diverges dramatically from its counterpart in ancient Greece. As we have seen, in ancient Greece the Pythagorean theorem led to the discovery of irrational numbers, such as $\sqrt{2}$. Chinese mathematicians, on the other hand, never explicitly formulated the concept of irrationality. Some historians have attributed the different paths of Chinese and Greek mathematics to the greater "practicality" of the former, and have suggested that the Chinese were not so interested in abstract reasoning. However, Liu Hui had no aversion to abstract reasoning or to the "impractical" side of mathematics. Joseph Dauben, a leading historian of Chinese mathematics, believes that the explanation lies in the Chinese language itself, in which it is difficult to express a counter-factual assumption. Recall that the proof of irrationality of $\sqrt{2}$ begins with the counter-factual sentence, "Let's assume that you could find a ratio p/q whose square is 2." Presumably, an ancient Chinese mathematician would simply not have been able to make sense of this first step. How can a false assertion be used to justify a true theorem?

Such differences are, once again, a reminder that there is no unique correct approach to mathematics. Even in the twentieth century, a school of mathematics called constructivism maintained that proofs by contradiction should not be allowed. They would find the argument above for the irrationality of $\sqrt{2}$ to be completely unconvincing. Like the ancient Chinese, they would be more interested in the fact that $\sqrt{2}$ is *computable*. (In other words, there is a well-defined procedure for approximating it to any desired degree of precision.) The more things change, the more they stay the same.

4
the circle game
the discovery of π

Besides computing the hypotenuse of a triangle, two other geometric problems seem to arise almost inevitably in any numerate civilization: finding the circumference and area of a circle. In contemporary notation, they are given by the closely related formulas $C = 2\pi r$ and $A = \pi r^2$. Here A represents the area of a circle of radius r, and π (read "pi") is the most famous constant in mathematics, the number 3.1415926535…

The modern formulas tend to obscure the first wonderful fact about pi: the fact that the same constant appears in both formulas. They obscure the fact by making it too obvious. To appreciate what ancient mathematicians had to figure out, we should imagine that there is a number "pi-circumference" defined by the ratio of a circle's perimeter to its diameter d, and a second number "pi-area" defined by the ratio of a circle's area to its radius squared. Imagine that you don't know that these two numbers are equal.

The first completely clear statement that the two problems are related comes, not surprisingly, in ancient Greek mathematics. In the third century BC, Archimedes wrote in his manuscript *The Measurement of the Circle*:

> **Proposition 1**. The area of any circle is equal to the area of a right triangle in which one of the sides about the right triangle is equal to the radius, and the other to the circumference of the circle.

$$\pi = 3.14159\ 26535\ldots$$

The irrational number π actually has two different meanings. First, it is the ratio of the area A of any circle to the square of its radius r. (That is, $\pi = {}^A/_{r^2}$.) Second, it is the ratio of the circumference **C** of any circle to its diameter, d. (That is, $\pi = {}^C/_d = {}^C/_{2r}$.) Either one of these statements may be taken as a definition of π, and then the other statement becomes a theorem.

Imagine cutting the circle into many wedges, each one indistinguishable from a triangle, so its area is half the base times the height of the wedge, as in the drawing on the next page. The height of each wedge is the radius of the circle, and the sum of all the bases is (roughly) the circumference of the circle. Thus the combined area of all the wedges is (roughly) half the radius times the circumference, and is also (roughly) equal to the area of the circle. The hard part of Archimedes' argument was turning the rough equalities into exact equalities. Once this is done, it is fairly easy to show that "pi-circumference" is the same as "pi-area."

Archimedes' Proposition 1 has been overshadowed by his Proposition 3, where he proved that π lies between $3\ ^1/_7$ and $3\ ^{10}/_{71}$. But it is really Proposition 1 that gives birth to the concept of pi. Without it, you have two separate problems: how to compute areas and circumferences of circles. With it, you can replace them with a single problem: how to approximate the number pi. Proposition 3 is merely an elaboration of that theme.

As in the case of the Pythagorean theorem, ancient Chinese mathematicians were not far behind their Greek counterparts, if at all. Already in the *Nine Chapters*—which may predate Archimedes—we find the following problem: "Given a circular field, the circumference is 181 *bu*

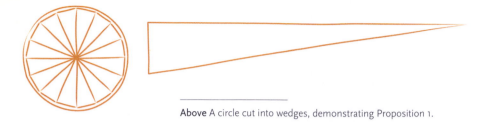

Above A circle cut into wedges, demonstrating Proposition 1.

and the diameter 60 ⅓ *bu*. Tell: what is the area? ... Rule: Multiplying half the circumference by the radius yields the area of the circle in [square] *bu*."

The third sentence (the "Rule") is nothing more or less than Archimedes' Proposition 1. Interestingly, the first sentence shows that the anonymous author thought that π = 3, a very primitive approximation.

HOWEVER, LIU HUI, the third-century commentator on the *Nine Chapters*, had other ideas. To start with, he pointed out that the ratio of the perimeter of a hexagon to its diameter is equal to 3, and yet the perimeter of a circle is visibly larger than that of a hexagon. So the ancient method, based on π = 3, could not be right. "The difference between a polygon and a circle is just like that between the bow and its chord, which can never coincide." Liu wrote. "Yet such a tradition has been passed down from generation to generation and no one cares to check it."

To compute a more precise "circle rate," his term for pi, Liu pushed out each side of the hexagon, to create a 12-sided polygon, and computed the perimeter of that. Then he repeated this procedure to obtain the perimeter of a 24-sided figure, a 48-sided figure, and a 96-sided figure. He also did the same procedure for a 12-sided polygon drawn outside the circle, a 24-sided polygon drawn outside the circle, and so on. In this way, he shows that:

$$314\frac{64}{125} < 100\pi < 314\frac{169}{125} \text{ or } 3.1408 < \pi < 3.1420$$

This is very comparable to Archimedes' estimate:

$$3\frac{10}{71} < \pi < 3\frac{1}{7} \text{ or } 3.1407 < \pi < 3.1428$$

Above Liu Hui's demonstration of his "circle rate."

Archimedes computed his estimate in exactly the same way—starting with a hexagon and doubling the number of sides until he got to a 96-gon! It is amazing that these two great minds, separated by so many miles and so many years, came up with exactly the same idea. The only reason for the slight difference in their answers is that Liu has made more careful approximations along the way. (After the first step, the perimeters involve square roots, which had to be approximated by rational numbers.)

But Liu, unlike Archimedes, didn't stop! He adds that the procedure can be continued all the way up to a 3072-sided polygon. He omits the calculations, but gives us the result:

$$\pi \approx \frac{3927}{1250} \text{ or } \pi \approx 3.1416.$$

He has gotten four digits of pi correct! Liu was probably the first human being to find this now standard approximation to pi.

Curiously, Liu worked pi out to this accuracy once, but in all of his annotations to the other problems in the *Nine Chapters*, he employed the simpler approximation $\pi \approx 3.14$. This inconsistency points out something very interesting about Liu's own psychology. He must have realized that there would be no conceivable use for a more accurate approximation in any practical problem. Unless you have a laser interferometer (which didn't exist back then), you can't measure the diameter of a field to four decimal places, and so there is no point in using a "circle rate" with that degree of accuracy.

And yet he worked it out to four decimals anyway! He didn't need to do it; he just wanted to satisfy his own curiosity. He was only the first of many math geeks (or perhaps more specifically pi geeks) over the centuries, who have pushed the computation of pi to almost unfathomable lengths. Before the computer era, William Shanks computed 707 digits of pi, although he tragically made a mistake on the 527th digit, and all the later digits were wrong. Now, with the advent of computers, the record number of digits has been pushed beyond one trillion.

TO PENETRATE this deep into the mysteries of pi, you need more than the relatively clumsy geometric approach of Archimedes and Liu. Around 1500, an unknown Indian mathematician of the Kerala school (possibly Nilakantha Somayaji or his predecessor Madhava) discovered the exquisite formula:

$$\frac{\pi}{4} = 1 - \frac{1}{3} + \frac{1}{5} - \frac{1}{7} + \frac{1}{9} - \dots$$

now known as the Gregory–Leibniz formula after its first European discoverers. Such formulas, relating pi to infinite sums of simple fractions, became much easier to derive with the invention of calculus by Isaac Newton and Gottfried Wilhelm Leibniz in the late 1600s. A personal favorite, proved by Leonhard Euler in 1734, is the amazing equation:

$$\frac{\pi^2}{6} = 1 + \frac{1}{2^2} + \frac{1}{3^2} + \frac{1}{4^2} + \dots$$

The symbol π for the "circle rate" was also introduced about this time (in 1706 by William Jones), and popularized by Euler.

Take a moment to reflect on the beauty of these formulas. These equations reveal that the number pi is not merely a geometric concept. Three of the great tributaries of mathematics merge in these formulas: geometry (the number pi), arithmetic (the sequence of odd numbers, and the sequence of squares $1^2, 2^2, 3^2, \dots$), and analysis of the infinite (in this case, infinite sums). Archimedes would have been flabbergasted to see formulas like these. Liu would have been speechless. And they would have gone straight out to buy a book on calculus and learn this wonderful new art.

And yet there are even deeper levels to the number pi. It is irrational—a fact that eluded ancient mathematicians, although they must have suspected

it. Johann Lambert proved the irrationality of pi in 1761. A century later, Ferdinand Lindemann (in 1882) proved the more subtle fact that pi is transcendental, a kind of souped-up version of irrationality.[‡] Lindemann's theorem resolved the ancient problem of squaring the circle, posed by the ancient Greeks: Is it possible, with only basic geometric operations, to draw a square whose area is the same as a given circle? A positive answer to this question—a method for squaring the circle—would have made the number pi more accessible to them. Alas, it was not to be. Transcendental ratios cannot be constructed with a ruler and compasses.

Even today, there are facts we still do not know about pi, and discoveries presumably still waiting to be made. As recently as 1995, three mathematicians—David Bailey, Peter Borwein, and David Plouffe—discovered a brand-new formula for pi that may deserve to be etched on the same mountaintop as Leibniz's and Euler's. It is the first self-repairing formula for pi, in the sense that if you make a mistake at the 527th place, it doesn't invalidate your later calculations. However, there is a catch. The self-correcting property is only true if you write pi in hexadecimal (base 16) arithmetic, as computers do.[§] It won't work in ordinary (decimal) notation. So if God created the integers, and God created pi, then perhaps God is actually a computer.

[‡] A number is transcendental if it cannot be expressed as the solution to any polynomial equation with rational coefficients. For instance, $\sqrt{2}$ is not transcendental, because it solves the equation $x^2 = 2$.

[§] In hexadecimal notation, π = 3.243F6A8885A308D3... The letters "A" through "F" stand for the numbers 10 through 15, which are single digits in base 16.

5

from zeno's paradoxes
to the idea of infinity

The city of Elea, in the present-day province of Salerno, Italy, was home to two noted philosophers who spanned the period between Pythagoras and Socrates. Parmenides, the elder of the two, was noted for his beliefs that all things are one and that the world we perceive is different from the world of reality—a viewpoint that would strongly influence Plato's philosophy.

Parmenides' student, Zeno, is noted not so much for any particular beliefs as for a style of debate that Aristotle called dialectic, in which you argue against your opponent's beliefs rather than in favor of your own. Zeno would take his opponent's belief as a premise and try to prove logically that the premise led to an absurdity. His arguments are usually called "paradoxes" because they seem to refute very commonly-held beliefs. For example, suppose that you believe that it is possible to move from point A to point B. Before you can reach point B, Zeno argues, you must have gone halfway to B. Before you can get halfway to B, you must have gone half of that distance (or a quarter of the distance to B), and so on. In other words, you must have completed an infinite number of motions before you can even travel the tiniest fraction of the distance from point A to point B! Clearly, Zeno says, this is absurd. Therefore, motion is impossible.

A second paradox is called Achilles and the tortoise. If you believe in motion, Zeno says, then you must surely believe that the swift Achilles can catch up with a slow-moving tortoise. But he argued that if Achilles runs to

$$1 + \tfrac{1}{2} + \tfrac{1}{4} + \tfrac{1}{8} + \dots = 2$$

The ellipsis (...) means that the sum is to be taken *ad infinitum*, not stopped after a finite number of steps. More formally, you can get as close as you want to 2 if you add up a sufficiently large (finite) number of terms in this sum.

where the tortoise is now, the tortoise will have already moved a few steps forward. If Achilles runs to that place, the tortoise will have moved forward once again, and so on. Again, Achilles must complete an infinite number of tasks in a finite time, and that (at least to Zeno) is clearly absurd.

To the modern-day mathematician, Zeno's paradoxes are harmless. In fact, they are a rather perceptive description of what continuous motion is all about. Let's say that Achilles is going twice as fast as the tortoise, and the tortoise starts with a 1-yard head start. After 1 second, Achilles has traveled 1 yard and the tortoise has traveled ½ yard. (This is one fast tortoise!) After 1 + ½ seconds, Achilles has traveled (1 + ½) yards and the tortoise has traveled (½ + ¼) yards. Where will Achilles and the tortoise be after n of these steps? And how much time will elapse? Working through the sums, the amount of time elapsed is just shy of 2 seconds. In fact, it is:

$$\left(1 + \frac{1}{2} + \frac{1}{4} + \dots + \frac{1}{2^n}\right) \text{ seconds, or more simply } \left(2 - \frac{1}{2^n}\right) \text{ seconds}$$

Achillles has traveled just less than 2 yards, in fact $\left(2 - \dfrac{1}{2^n}\right)$ yards

The tortoise has traveled only half that far, i.e. $\left(1 - \dfrac{1}{2^{n+1}}\right)$ yards

But it had a 1-yard head start, and if we add its head start to the distance it has traveled, we see it is $(2 - \dfrac{1}{2^{n+1}})$ yards ahead of Achilles' starting point.

Because $\dfrac{1}{2^{n+1}} < \dfrac{1}{2^n}$, the tortoise is closer to the 2-yard mark than Achilles is.

Therefore Zeno is correct in asserting that *after* $(2 - \dfrac{1}{2^n})$ seconds Achilles is still behind the tortoise.

So we know where Achilles and the tortoise are just a split second before 2 seconds, and a split split second after that, and a split split split second after *that* … But no man, not even Zeno, can stop time. Eventually the stopwatch will reach 2 seconds. Where will Achilles and the tortoise be then? The answer is that they will both be at the point they have been getting closer and closer to—2 yards beyond Achilles' starting point. Modern mathematicians call this "taking the limit" as n approaches infinity. Both of the terms $1/2^n$ and $1/2^{n+1}$ approach zero, and so disappear in the limit. After 2 seconds, Achilles has traveled 2 yards, the tortoise has traveled 1 yard; Achilles has caught up.

Where did Zeno go wrong? First, he started to mathematize the problem but did not finish the job: he left out important information, namely the time elapsed. Second, and more importantly, he and the other ancient Greeks were still sufficiently uneasy about the concept of infinity that they could not take the limit. That is, they could not go from the finite sum:

$$1 + \frac{1}{2} + \frac{1}{4} + \ldots + \frac{1}{2^n} = 2 - \frac{1}{2^n}$$

to the infinite sum:

$$1 + \frac{1}{2} + \frac{1}{4} + \ldots = 2$$

But they tried so hard! And they came so close! Just how close becomes apparent when you read Archimedes' *Quadrature of the Parabola*, written about two centuries after Zeno.

In this document, written as a letter to a fellow mathematician, Dositheus, upon the death of a mutual friend named Conon, Archimedes writes: "While I grieved for the loss not only of a friend but of an admirable mathematician,

I set myself the task of communicating to you, as I had intended to send to Conon, a certain geometrical theorem which had not been investigated before but has now been investigated by me, and which I first discovered by means of mechanics and then exhibited by means of geometry. Now some of the earlier geometers tried to prove it possible to find a rectilineal area equal to a given circle … but I am not aware that any one of my predecessors has attempted to square the segment bounded by a straight line and a section of a right-angled cone [i.e., a parabola]." He goes on to say that he has proved that any such region has area equal to $4/3$ times the area of an inscribed triangle whose height is the same as the height of the parabolic region.

Quotes such as this give us insight into the character of Archimedes—another math geek. The best way he can think to console Conon's friend is to send him the proof of a new mathematical theorem! Also, notice the reference to squaring the circle—a subject that Archimedes had some experience with.

Archimedes is not able to square the circle but he is able to "square" or rectify a different curved region, a far from obvious accomplishment. And finally, notice he draws a curious distinction between "discovering" the theorem by means of mechanics and "exhibiting" (or proving) it by means of geometry.

AS IT TURNS OUT, the method that Archimedes used to estimate the area of a circle works much better for a parabola. And this time there is nothing approximate about it: Archimedes says that the area of the parabola is exactly $4/3$ that of the inscribed triangle. To prove this, he pushes out two of the sides of the triangle T, creating a four-sided figure that more closely approximates the parabola. Then he pushes out those sides, creating an eight-

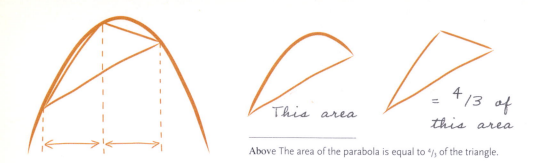

This area = 4/3 of this area

Above The area of the parabola is equal to $^4/_3$ of the triangle.

sided figure, and so on. And at each step, he shows, he adds one quarter of the area that was added in the previous step. So if we take the area of the initial triangle as 1, then the area of the four-sided figure is $1 + ^1/_4$. The area of the eight-sided figure is $1 + ^1/_4 + ^1/_{16}$. Continuing in this fashion, after n steps, he has an extremely close approximation to the parabola, whose total area is:

$$1 + \frac{1}{4} + \frac{1}{4^2} + \dots + \frac{1}{4^n}$$

This strongly resembles the sum seen when analyzing Zeno's paradox, only involving powers of 4 rather than 2. Next, Archimedes shows that adding $^1/_3$ of the last term to this finite sum always gives a total of exactly $^4/_3$:

$$1 + \frac{1}{4} + \frac{1}{4^2} + \dots + \frac{1}{4^n} + \frac{1}{3}\left(\frac{1}{4^n}\right) = \frac{4}{3}$$

Remember that the ancient Greeks were uncomfortable with numerical proofs, so Archimedes had to prove this geometrically, as shown below. Suppose the L-shaped region labeled A has area 1. Then the area of the large square containing it is $^4/_3$ (because the large square has four equal quadrants, only three of which are contained in region A). The large square can be filled out, or "exhausted," by a shrinking sequence of L-shaped regions, plus one small leftover square in the lower right-hand corner

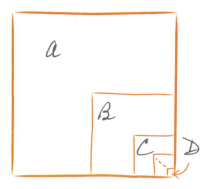

Left The area of the large square is $^4/_3$ the area of the region labeled A.

(labeled D). The total area of these pieces is the left-hand side of the equation above. Thus the left-hand side and the right-hand side are equal.

What a wonderful argument! But notice that Archimedes stops after n steps; he doesn't let the procedure run "on to infinity." A modern mathematician would have no qualms about taking this step. The parabola is "exhausted" by successive triangles, just as the square is exhausted by the successive L-shaped figures. Taking the limit as the number n approaches infinity, we would conclude that the parabola's area equals the square's area, which we showed was $4/3$. In other words, we would use the infinite sum:

$$1 + \frac{1}{4} + \frac{1}{4^2} + \ldots = \frac{4}{3}$$

This sum is appropriately called a geometric series in honor of its geometric origins. The terms in a geometric series decrease by a constant ratio from one term to the next. In Zeno's case the ratio was $1/2$; in Archimedes' case it was $1/4$. The general rule, as you might have guessed, is this:

$$1 + \frac{1}{r} + \frac{1}{r^2} + \ldots = \frac{r}{r-1}$$

Unfortunately, the mathematics of Archimedes' time would not allow him to take this final step. Instead, he had to resort to an ingenious *reductio ad absurdum* argument. Just as Zeno would have done, he argues against an imaginary opponent. You think that the parabola's area doesn't equal $4/3$? Fine. Then you must tell him whether it is greater than $4/3$ or less than $4/3$. If you say it's greater than $4/3$, then Archimedes will show by his subdivision technique that you have overestimated the area. If you say it's less than $4/3$, he will show that you have underestimated. Either way you lose, and you have to concede that the area is $4/3$.

With hindsight, we can see that Archimedes has come a long, long way towards understanding infinite processes[*]. He is using the infinite to discover new truths—a huge leap forward, which took the ancient Greeks to the brink of mastering the infinite.

---

[*] Actually, Archimedes was far from alone. Eudoxus of Cnidos (408–355 BC) is credited with inventing the "method of exhaustion," which Archimedes employed to such good effect here.

6

a matter of leverage

laws of levers

Archimedes of Syracuse was born about 287 BC and died in 212 BC. In his own time, his reputation rested more on his physical discoveries and engineering inventions than on his mathematics—yet he would surely have considered himself a mathematician. He was proudest of his proof that a sphere has two-thirds the volume of its circumscribed cylinder; or equivalently, $V = (4/3)\pi r^3$. He even asked for a diagram of a sphere and a cylinder to be inscribed on his tombstone.

Archimedes' work represents a beautiful unification of applied mathematics with geometry and with the still gestating concept of the infinite. Today, however, most people probably associate Archimedes with the story that he ran down the road naked, crying "Eureka!" ("I have found it!") The story goes that Archimedes' friend and patron, King Hieron, wanted to know if a certain crown was made of pure gold. Archimedes supposedly was sitting in his bathtub when the solution occurred to him. If the crown was made of a cheaper alloy, it should be less dense than pure gold. If placed in a container of water, the ersatz crown would displace more water than a gold piece of the same weight.

Often, legends like this are codified versions of real events. One example was the purported drowning of Hippasus by the Pythagoreans, and another will be

$$d_1 w_1 = d_2 w_2$$

In a lever, a weight w_1 at distance d_1 from the fulcrum will balance a weight w_2 at distance d_2.

seen in the discussion of how Isaac Newton discovered the law of gravity. In reality, Archimedes wrote a book called *On Floating Bodies*, in which he formulates what became known as Archimedes' principle: the weight of water displaced by an object (either floating or completely immersed) equals the buoyant force exerted by the water on the object.

From Archimedes' principle it is possible to deduce a formula for the density of an object immersed in water. If ρ_{body} denotes the density of the object, ρ_{fluid} denotes the density of the fluid (which, in the case of water, is conventionally assumed to be 1), w_{dry} denotes the dry weight of the object and $w_{immersed}$ denotes its apparent weight when fully immersed, then:

$$\rho_{body}/\rho_{fluid} = w_{dry}/(w_{dry} - w_{immersed})$$

This formula makes it possible to compute the crown's density (or "specific gravity") directly, so there would be no need to compare it with an actual gold crown of equal weight. Archimedes surely came to this discovery over a period of time, not overnight, and it is pretty certain that he was aware of it before King Hieron came to him. Everything else about the legend is embellishment, including running naked through the streets.

Archimedes' principle and the specific gravity formula are still used routinely, even today. The rest of *On Floating Bodies* contained a wealth of information about bodies of different shapes and their stable floating configurations. It was a first step toward making shipbuilding a science instead of a matter of trial and error.

Archimedes also experimented with levers and pulleys. Again, there is a story that King Hieron remarked on the power of Archimedes' contraptions, and Archimedes replied, "Give me a place to stand and I will move the Earth." Archimedes understood the lever law, $d_1w_1 = d_2w_2$, which expresses the relationship between the weights of two objects (w_1 and w_2), if they are balanced on a lever at distances d_1 and d_2 from the fulcrum. From the formula it follows that a lesser weight can balance a greater weight if it is farther from the fulcrum. For instance, a 150-pound man can lift a 1500-pound safe if he stands on one end of a lever, places the safe at the other end, and places the fulcrum at least ten times closer to the safe than to himself.

Archimedes frequently used the lever law not only in physical devices, but also in mathematical research. As discussed earlier, he first discovered the area of a parabola "by means of mechanics" and only later proved it "by

Below "Give me a lever and I will move the Earth" – a woodcut showing Archimedes putting his famous saying into action from the title page of *The Mechanic's Magazine* London, 1824.

means of geometry." That argument, filling the parabolic segment up with triangles, was actually his second proof. His original proof involved an equally ingenious and different way of cutting up the parabolic segment into pieces, and balancing those pieces with rectangles of known area (or weight) on the other side of a lever. It was actually a favorite method of his. However, Archimedes apparently felt that the lever law was too informal or perhaps too empirical to be acceptable as pure mathematics. Thus, after "discovering" a theorem with levers, he felt compelled to confirm it in a way that Euclid would have approved.

ARCHIMEDES WAS FORTUNATE enough to live most of his life during the prosperous and peaceful 54-year reign of King Hieron. Unfortunately, toward the end of his life that period of peace came to an end. Hieron's son antagonized the growing Roman Empire, and the result was a one-sided war between the Romans and the Syracusans.

Almost single-handedly Archimedes was able to hold off the Roman army, by designing grappling devices and cranes of unprecedented accuracy. According to Plutarch, the Romans became so terrified of Archimedes' devices that "if they only saw a rope or a piece of wood extending beyond the walls, they took flight exclaiming that Archimedes had once again invented a new machine for their destruction."

As a last resort the Roman general, Marcellus, laid siege to Syracuse. After two years the Romans entered the city. Marcellus gave orders to spare the life of Archimedes, but, according to legend, a soldier came upon Archimedes kneeling over a mathematical diagram. "Don't disturb my circles," Archimedes told him. Enraged by the unknown man's impudence, the soldier ran him through with his sword.

equations in the age of exploration

On August 10, 1548,

the Church of Santa Maria del Giardino in Milan, Italy, was thronged by curious spectators. The event they came to witness was not a church service, but the mathematical equivalent of a duel at twenty paces. Using nothing but their wits, Niccolò Tartaglia of Venice would battle against Lodovico Ferrari, a peasant boy-turned-servant to one of Milan's most famous citizens: Girolamo Cardano—a physician, gambler, and jack of all intellectual trades.

Curiously, Cardano himself was nowhere to be found. He had precipitated the ruckus three years earlier by publishing a mathematical formula that Tartaglia had given to him in strictest confidence. However, on this day he had found a convenient excuse to be out of town while his servant, who was in all likelihood a better mathematician than himself, defended his honor.

The competitors must have seemed better suited for a back-alley brawl than a contest of minds. Tartaglia had been disfigured as a youth by a deep saber wound to his jaw, received when a French army sacked his hometown of Brescia in 1512. Though as an adult he hid his scar by growing a full beard, the injury had left him with a permanent speech defect that led to his nickname: Tartaglia, the stammerer. Ferrari, too, bore the scars of a rough-and-tumble childhood, as he was missing some of the fingers of his right hand.

We will never know exactly what happened in the church that day. The planned contest of minds apparently turned into a shouting match. But circumstantial evidence suggests that Ferrari was perceived as the winner. The governor of Milan, who was in the audience, was impressed enough by Ferrari's talent to hire him as a tax assessor. Tartaglia lost a teaching position he had just gotten in Brescia, for which he never received a penny. He died nine years later as Cardano's sworn enemy. As for the man who had started it all, Cardano returned home with his reputation intact and continued to enjoy the life of the proverbial Renaissance man.

The battle in the church was the final act of one of the most bitter and bizarre disputes in the history of mathematics, a debate over the rights to the first completely new mathematical discovery in Europe since the fall of the Roman Empire. Until the early 1500s, western Europe had mostly been playing catch-up to the rest of the world, as well as to its own past. The formula that Tartaglia had confided to Cardano—which is now known, rather unjustly, as Cardano's formula—has been compared to the discovery of America, because it was a new fact about the world that was not even hinted at in any ancient books. It launched an Age of Exploration in mathematics that would transform the map of the mathematical world as profoundly as Columbus's discovery transformed the map of the physical world.

7

the stammerer's secret
cardano's formula

The story of Cardano's formula really begins more than 3000 years earlier. In the period between 1850 and 1650 BC, problems like this one proliferated in Babylonian mathematical tablets: find two numbers whose product is 60, and whose difference is 7. A modern mathematician would call the numbers x and y and note that $y = 60/x$. Therefore, $x - 60/x = 7$, or equivalently, $x^2 - 7x - 60 = 0$. Then one would trot out the *quadratic formula*, which says that the solution to any quadratic equation, $ax^2 + bx + c = 0$, is given by:

$$x = \frac{-b \pm \sqrt{b^2 - 4ac}}{2a}$$

Using $a = 1$, $b = -7$, and $c = -60$, said mathematician would obtain the solutions $x = 12$ and $y = 60/12 = 5$.

However, the Babylonians did not have the algebraic tools that we do today. Instead, the scribe took a more intuitive approach, which involved drawing a rectangle with the unknown side lengths, x and y, cutting it into pieces, and rearranging the pieces into an L-shaped figure, as shown on page 62. He then "completed" the L-shaped figure by adding a small square of known area in the corner. A similar method of solution—called "completing

$$x = \sqrt[3]{\frac{q}{2} + \sqrt{\frac{q^2}{4} + \frac{p^3}{27}}} - \sqrt[3]{-\frac{q}{2} + \sqrt{\frac{q^2}{4} + \frac{p^3}{27}}}$$

Cardano's formula for solving a reduced cubic polynomial, $x^3 + px = q$

the square"—is still taught in high-school algebra as a precursor to the quadratic formula, but usually with no reference to its geometric meaning or its historical provenance.

The other ancient mathematical cultures also "knew" the quadratic formula or else had equivalent methods for solving quadratic equations. Euclid employed a geometrical construction that produced a line segment of the requisite length. In seventh-century India, Brahmagupta, discussed previously in relation to zero, provided a recipe to solve the equation $ax^2 + by = c$ that is essentially the quadratic formula written in words instead of symbols.

However, classical mathematics is essentially silent[*] on the question of how to solve a *cubic* equation, $ax^3 + bx^2 + cx + d = 0$. In 1494, Fra Luca Pacioli, an Italian mathematician, expressed the opinion that cubic equations would *never* be solved exactly. Pacioli was proved wrong only a generation later!

---

[*] An exception is the Persian poet and mathematician Omar Khayyam (1050–1130), who showed how to solve a large class of cubic equations by geometrical constructions (e.g., by finding the intersection point of a parabola and a circle). However, the results are not readily convertible into numerical form, and are not equivalent to the later work of Tartaglia and Cardano.

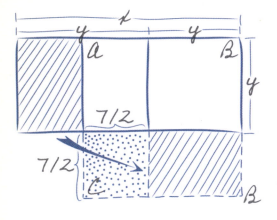

In the early 1500s, a Bolognese mathematician named Scipio del Ferro apparently found a method for solving cubic equations that are lacking the quadratic term: in other words, any equation of the form $x^3 + px = q$. Nowadays, a mathematician who made such a discovery would hasten to publish it. However, in the Italy of that era, mathematicians made their reputations by defeating other mathematicians in problem competitions. Del Ferro therefore kept his method secret, so that he could pose problems that his opponents would not be able to solve. Only on his deathbed did he confide his secret to two of his students, Antonio Maria Fiore and Annibale della Nave.

However, rumors soon spread about del Ferro's discovery. In the early 1530s, Tartaglia started claiming that he, too, could solve cubic equations. Thinking he could call Tartaglia's bluff, Fiore rashly challenged Tartaglia to a competition. According to the legend (which is probably a little too good to be true), on the eve of the debate Tartaglia finally figured out how to solve these cubics, and so he thoroughly trounced Fiore.

How, then, did Cardano get his name on what should have become known as Tartaglia's or del Ferro's formula? It becomes a little less surprising when you read Girolamo Cardano's own description of himself: He wrote in his autobiography that he was "hot tempered, single minded, and given to women … cunning, crafty, sarcastic, diligent, impertinent, sad and treacherous, miserable, hateful, lascivious, obscene, lying, obsequious …" He had studied to become a physician at the University of Padua, but perhaps because of his erratic behavior he was forbidden to practice medicine in Milan until 1539. However, even before then he found great success as a public lecturer and writer on a variety of topics, including mathematics.

In 1539 Cardano was composing a mathematical handbook called *Practica arithmeticae generalis*, and he asked Tartaglia for the secret to solving cubic equations. Tartaglia at first refused, on the grounds that he intended to write a book of his own. But at last, Cardano persuaded Tartaglia to come to his house for a visit. On the night of March 25, 1539, Tartaglia revealed his method under an oath of strict secrecy.

And here is the secret, in an English translation by math historian Jacqueline Stedall. (Tartaglia's version, in Italian, actually rhymes!) For ease of understanding, Tartaglia's verse has been interpreted here in algebraic symbols. "The thing" means the unknown quantity x, "the cube" means x^3, and "the number of things" is p. The equation Tartaglia wants to solve is $x^3 + px = q$, and the numbers p and q are positive. (This last point is irrelevant to mathematicians today, but was very relevant to sixteenth-century Italians, who were still as skeptical of negative numbers as the nineteenth-century-BC Babylonians.)

Tartaglia	Algebra
When the cube with the things next after	When $x^3 + px$
Together equal some number apart Find two others that by this differ	$= q$, Find u and v such that $u - v = q$,
And this you will then keep as a rule	
That their product will always be equal	And such that $uv =$
To a third cubed of the number of things	$(\frac{p}{3})^3$
The difference then in general between	Then
The sides of the cubes subtracted well	$\sqrt[3]{u} - \sqrt[3]{v}$
Will be your principal thing.	$= x$

Tartaglia's formula is a well-disguised way of "completing the cube," but it includes a very clever new step that was not present in the Babylonian process of completing the square: the introduction of two new auxiliary variables, u and v. Here is how it works for an example considered later by Cardano: $x^3 + 6x = 20$. Tartaglia instructs us to find two numbers u and v such that $uv = (6/3)^3 = 8$ and $u - v = 20$. This pair of equations can be solved by the quadratic formula: $u = 10 + 6\sqrt{3}$ and $v = -10 + 6\sqrt{3}$. (These are the two positive solutions.) Now we are supposed to find the cube roots of u and *v*. In general one would have to do this by approximation, either by hand or with an abacus. But in this particular case the cube roots have a simple exact form:

$$\sqrt[3]{u} = \sqrt{3} + 1 \text{ and } \sqrt[3]{v} = \sqrt{3} - 1$$

Finally, we subtract these to get the answer: $x = (\sqrt{3} + 1) - (\sqrt{3} - 1) = 2$

Below An engraving from frontispiece of Cardano's work *Ars Magna*, 1545, the first great Latin book dedicated to algebra.

AT FIRST, Cardano honored his pledge not to publish Tartaglia's method. However, over the next few years several things happened that made him itch to see the solution in print. First, he and his protégé, Ferrari, went beyond Tartaglia, by showing how to simplify any cubic equation to del Ferro's form or one of 12 other basic forms. Second, Ferrari "invented at my request" (as Cardano later wrote) a method for solving quartic or fourth-degree equations. This latter discovery is far more remarkable than Cardano's offhand comment would suggest. More than 3000 years elapsed between the solution of the quadratic and the first solution of the cubic—but it took Ferrari only four years to move on to quartics! Unfortunately, the solution to the cubic was an intermediate step in the solution of quartics. The promise to Tartaglia was now a major obstacle: Without the method for solving cubics, Cardano could not publish Ferrari's brilliant solution for quartics.

At this point Cardano found an ingenious loophole. In 1543,

he tracked down del Ferro's other student, della Nave in Bologna, and ascertained that Tartaglia's method for solving the cubic was exactly the same as del Ferro's. This fact apparently released Cardano (at least, in his own mind) from his promise to Tartaglia to keep it a secret. Two years later, Cardano published his greatest mathematical work, *Ars magna* (*The great art*), with a complete treatment of cubics and quartics, and the secret was out.

Tartaglia, of course, felt that Cardano had betrayed him. He fired off a volley of insulting open letters, as well as a book of his own. Cardano, however, remained above the fray and allowed Ferrari to do the answering for him—a task that Ferrari took to with great zest, and ultimately with the successful outcome recounted at the beginning of this chapter.

It may seem unjust that the formula for solving the cubic is now known as Cardano's formula—not del Ferro's, or Tartaglia's, or even Ferrari's. But as has already been stated in the last chapter, mathematics thrives when it is communicated openly. It is not enough merely to discover America—you must make the discovery known to the rest of the world. Cardano alone took that final step, and reaped the glory.

CARDANO'S FORMULA had a lasting impact that far exceeded the importance of the problem it solved. For example, it provided one of the first motivations for the use of imaginary numbers and complex numbers in mathematics. Imaginary numbers are numbers whose square is negative (a property no real number has). Using imaginary numbers, we can say that −1 has two square roots, which are denoted by i and $−i$. Without imaginary numbers, we would have to say that −1 has no square roots. Once we have imaginary numbers, we can define complex numbers as numbers that have both a real and an imaginary part, such as $1 + 2i$.

Not only modern-day mathematics but also modern-day physics would be unthinkable without imaginary numbers. In quantum mechanics, for instance, elementary particles such as photons are defined to be "wave functions." The wave function for a photon at some point will in general have complex-number values, such as $0.2 + 0.3i$. The imaginary part of the wave function accounts for the wavelike properties, or "phase," of the photon; for instance, it explains why a light beam that shines through two slits forms a

diffraction pattern on the other side, rather than two bright bars (see Young's experiment, on page 143). Thus, imaginary numbers seem to be woven in a very real way into the fabric of the universe.

Prior to Cardano, it did not occur to anybody to assert that −1 has two square roots but they are imaginary. You might compare it to asking a child how many imaginary friends she is inviting to her birthday party. But in the case of cubic polynomials, these "imaginary friends" actually left behind some real birthday presents! In 1572, Rafael Bombelli presented an example of this phenomenon. The equation $x^3 = 15x + 4$ has a real solution, $x = 4$, that can be verified by substitution. However, Cardano's formula gives:

$$x = \sqrt[3]{2 + \sqrt{-121}} + \sqrt[3]{2 - \sqrt{-121}}.$$

Cardano had no interest in such nonsense: "So progresses arithmetic subtlety, the end of which, as is said, is as refined as it is useless," he once wrote. But Bombelli realized that these expressions have meaning. The first cube root is equal to $2 + i$ and the second is equal to $2 - i$, and therefore $x = (2 + i) + (2 - i) = 4$. In the final solution the imaginary quantities have disappeared, but we could not have gotten to the solution without them.

Thankfully, today's students are no longer expected to learn Cardano's formula. Nevertheless, nineteenth-century students were expected to know how to solve cubics. Albert Einstein, in his university exams, correctly solved a problem with Cardano's formula—in contrast to the surprisingly persistent legend that he was a poor mathematics student.

Another long-delayed ramification of Cardano's formula involved the solution of higher-degree equations. After the cubic and quartic had been tamed, one might have expected the solution of fifth-degree polynomials, or quintics, to follow shortly thereafter. But strangely, another 250 years went by with very little progress. Some quintic equations can be solved. But no universal solution, applicable to all quintics, was ever found.

In 1824, a Norwegian mathematician named Niels Henrik Abel finally showed that there could not be any Cardano-like formula for the solutions to a fifth-degree equation. ("Cardano-like formula" refers to any formula that involves square roots, cube roots, fourth roots, etc., possibly nested inside one another. Mathematicians call this a "solution by radicals.") Abel's theorem may have closed one chapter of mathematics, but it opened another. His proof led mathematicians to a deeper understanding of the concept of symmetry, a topic to be discussed in chapter 14.

8

order in the heavens
kepler's laws of planetary motion

Another great battle of sixteenth- and seventeenth-century science was fought over a "revolutionary" theory that actually was not a revolution. In 1543, while on his deathbed, Nicolaus Copernicus published a book called *De revolutionibus orbium coelestium*, which placed the Sun, not Earth, at the center of the solar system. Although Copernicus' theory was at odds with the ecclesiastical understanding of the cosmos, it was definitely *not* a new idea. Aristarchus of Samos, a Greek philosopher, had already discussed a heliocentric model of the universe in the fourth century BC.

In the early years of the 1600s, two events thrust the "Copernican" (but really Aristarchan) theory into the center of a storm of controversy. The first was the invention of the telescope in 1608. Secondly, using one of these new instruments, Galileo Galilei discovered four small moons orbiting Jupiter. For Galileo, and for anyone else who took the trouble to look through the telescope, here was direct visual evidence of objects in the universe that did *not* orbit Earth. Galileo's discovery sounded the death knell for the dogma that Earth was the center of the universe.

It is easy to paint Galileo as the great champion of the Copernican theory, and indeed his story is full of drama and martyrdom. Brought to trial by the Inquisition in 1633 on a charge of heresy for advocating the view that the

$$r(\theta) = \frac{p}{1 + \varepsilon \cos(\theta)}$$

$$r^2(t)\,\frac{d\theta}{dt} = C_1$$

$$I = C_2 R^{3/2}$$

The function $r(\theta)$ represents the distance of a planet from the Sun when its location on the zodiac is θ degrees. The angular position $\theta(t)$ is itself a function of time, t. The total time it takes the planet to go around the Sun is T. The constants p and R describe (roughly) the width and length of the orbit. The eccentricity, ε, describes how far the orbit deviates from a perfect circle. C_1 and C_2 are two empirical constants.

Earth was not the center of the universe, Galileo was convicted, forced to recant, and confined to house arrest for the rest of his life. However, there is a second hero of the story who is not quite as well known, yet perhaps deserves equal credit: Johannes Kepler.

Although he is primarily considered an astronomer, Kepler had a real gift for mathematics and for bold conjecture. He was able to spot patterns where no one else had before—and sometimes where none existed. For example, when Galileo announced his discovery that Jupiter had four moons, Kepler conjectured that Mars must have two moons and Saturn eight, in order to make a geometrical progression: 1 (Earth), 2 (Mars), 4 (Jupiter), 8 (Saturn). Amazingly, he was right about Mars, but the "pattern" was an utter coincidence. Jupiter has 63 moons that we know about, and Saturn has 62.

It is easy to understand why Kepler, with his speculative temperament, was one of the first scientists to wholeheartedly embrace Copernicus' theory. It is a bit more of a surprise to see him chastising Galileo for not doing the same. In 1597, when Galileo wrote to him that he agreed with Copernicus but dared not publish his opinion, Kepler wrote back: "I would have wished, however, that you, possessed of such an excellent mind, took up a different position ... Have faith, Galilei, and come forward!" Nevertheless, Galileo remained publicly silent for thirteen more years, until his discovery of the Galilean moons of Jupiter gave him the evidence he needed to "come forward."

NOWADAYS, KEPLER'S FAME rests on three mathematical laws that he discovered by analyzing the painstaking observations of the orbits of the planets that had been taken by his former employer, the Danish astronomer Tycho Brahe. His laws form a bridge from the old style of astronomy, which was concerned with describing the cosmos, to a new style that explains the motions of the planets and other celestial bodies. They are still descriptive laws, but they are so precise that they virtually beg for a mathematical proof. Isaac Newton provided the proof in 1686, roughly three-quarters of a century later.

Kepler's first law states that planets orbit the Sun in ellipses, not circles, with the Sun at one focus. To express this in equation form, we could write it as follows:

$$r(\theta) = \frac{p}{1 + \varepsilon \, \cos(\theta)}$$

Here $r(\theta)$ represents the distance from the planet to the Sun when it is θ degrees away from aphelion (its greatest distance from the Sun). The number p represents the distance when the planet is 90 degrees away from aphelion, and the number ε represents the eccentricity, or departure from circularity, of the planet's orbit. Notice that if the eccentricity ε is zero, the equation becomes $r(\theta) = p$: in other words, the distance from the planet to the Sun is a constant. In this case, and only in this case, the orbit is a circle.

Ironically, Kepler's first law was actually a departure from strict Copernicanism. The Sun is not at the center of Earth's orbit, but slightly displaced from it. At its closest approach to the Sun, called perihelion, Earth is about 91.3 million miles (147 million kilometers) away. At its greatest distance, Earth is 94.5 million miles (152 million kilometers) away. More importantly, Kepler's law made a clean break with centuries of tradition that tried to describe planetary orbits either as circles, or as complicated combinations of circular motions. Aristotle had considered circles to be the

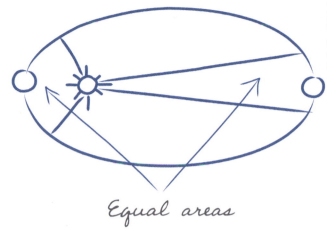

Equal areas

most perfect curves, and therefore the only ones that could describe the motion of the heavenly bodies. In the third century, Ptolemy had refined Aristotle's system with a complicated, Rube Goldberg-esque arrangement of circular motions superimposed on other circular motions—but even so he could not predict the planets' motions very accurately. Kepler's law (to modern eyes) is simple, economical, and far more beautiful than Plutarch's theory. To determine the shape of any planet's orbit, you need to know only two numbers: p and ε.

But Kepler wasn't finished. He discovered two more patterns in Brahe's planetary data that have since been elevated to the status of "laws." Kepler's second law states that planets speed up when they get closer to the Sun, and they do so in a precisely quantifiable way. The area that a planet sweeps out in any given, fixed-time interval is the same no matter where the planet is in its orbit. Because Earth is closer to the Sun at perihelion, it must sweep out a fatter triangle in one day at perihelion than it does when it is at aphelion, as shown above.

Earth's orbit is so nearly circular ($\varepsilon = 0.0167$), that most of us are not aware of these subtle differences. Nevertheless, they do affect us, more profoundly than most people realize. At present, Earth's closest approach to the Sun falls during the northern hemisphere's winter. This means that the northern hemisphere has somewhat shorter (and milder) winters than it would if the perihelion came in summertime. However, this will not always be the case. In about 13,000 years, the situation will be reversed, and we will experience

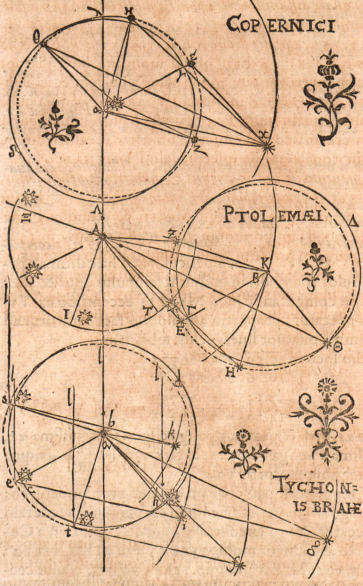

COPERNICI

PTOLEMÆI

TYCHON=
IS BRAHE

*strabitur (ut prius)
iisdem plane nu-
meris, lineis & an-
gulis, has lineas
præter opinionem
esse inæquales, ac
propterea Martem
non in circulo Γ Δ
versari, cujus sit
centrum in K pun-
cto æqualitatis mo-
tus, sed in ZEHΘ
circulo, cujus cen-
trum a K versus B
vergat, propemo-
dum in linea K B.
quæ sit parallelos
lineæ ex A TERRA
per perigæum So-
LIS ductæ.*
Vergit igitur a-
pogæum epicy-
cli in perigæum
SOLIS. Et quia e-
picyclus propter
omnimodam æ-
quipollentiam,
ut jam dictũ, po-
nendus est æqua-
lis circuitui Solis,
& Z K parallelos

ipsi ΞA, & EK ipsi OA, & HK ipsi IA, & ΘK ipsi TA: igitur etiam ipsas
ΞA, OA, IA, TA, inæquales esse verisimile est: & punctum medii loci
SOLIS (BRAHEANA notione centrum epicycli SOLIS) per circuitum a
puncto æqualitatis distare inæqualiter. Quod obiter interjeci. nihil .n.
facit ad præsentem demonstrationem, nisi quod eam extendit amplius.

　In forma TYCHONICA sit A TERRA, & ex ea scribatur SOLIS con-
centricus C D, qui putetur esse deferens SYSTEMA *Planetarum;* cum sit A
punctum æqualitatis motus concentrici SOLIS. Erit itaque SOL ipse in alio
eccentrico circulo. Sit ejus centrum ab A versus partes B. Sit autem A L re-
gula lineæ apsidum MARTIS, ut linea apsidum circulatione & transpositione
sui eccentrici semper maneat parallelos ipsi AL. Sint autem lineæ medii motus SOLIS ad nostra quatuor momenta AH, AT, AE, AS: & ex A ejician-
tur lineæ visionum MARTIS, prout supra descriptæ sunt, in hunc vel illum

longer and more severe winters. The existence of orbital variations like these has been proposed as a contributor to ice ages.

While Kepler's first two laws were the culmination of an eight-year struggle to understand Brahe's data on the orbit of Mars, Kepler's third law seems to have occurred to him quite suddenly—on March 8, 1618, as he was putting the finishing touches to a book called *Harmony of the World*. Unlike the first two laws, which describe the motion of an individual planet, the third law provides a basis for comparison between planets. It says that the length of a planet's year is proportionate to the 3/2 power of its distance to the Sun. (Another way of saying this is that the *square* of the orbital period is proportional to the *cube* of the mean distance from the Sun.) For instance, Pluto is 39.5 times farther from the Sun than Earth is. Thus it takes $(39.5)^{3/2}$ = 39.5 $\sqrt{39.5}$ = 248 years to orbit the Sun.

Kepler's third law is actually more useful in reverse. It is easy to measure the orbital time of a planet around the Sun, or a moon around a planet, or two stars around each other. The hard part is measuring the distance between them. Kepler's law gives us an immediate way of converting orbital periods to distances. Later improvements, using Newton's law of gravitation, enable us to infer the mass of moons, planets, or stars from their orbital periods. Such calculations are essential, for instance, in the search for extrasolar planets (i.e., planets in other solar systems) that might be capable of supporting life. If we didn't know how big and how far away from its sun a planet is, we would not know whether it is habitable. If, one day, we do find evidence of life on a distant planet, we will owe it to Kepler and his third law.

9

writing for eternity
fermat's last theorem

Pierre de Fermat was not a practical joker. The son of a wealthy leather merchant in southern France, he earned a law degree at the University of Orleans in 1631, bought a seat in the *parlement* (supreme court) at Toulouse, and became a member of the nobility. From the evidence of his letters, he was a shy, taciturn man who disliked controversy.

But Fermat had one unusual characteristic: He loved mathematics. In an era when mathematicians were starting to reach across national boundaries and turn their subject into an international enterprise, he achieved worldwide fame that lasted long after his death. By a curious twist of fate, his most lasting legacy was a problem that he almost certainly did not solve. That problem, called Fermat's Last Theorem, unintentionally became the greatest practical joke in mathematical history—a deceptively simple statement that defied all efforts at proof for more than 350 years.

To the best of our knowledge, Fermat was self-taught. However, during his student days he formed friendships with a small circle of people who were interested in mathematics, and this apparently stimulated him to start doing his own research. One of his friends moved to Paris in 1636 to work in the royal library, and brought the work of this previously unknown provincial mathematician, Fermat, to the attention of Father Marin Mersenne.

$$x^n + y^n = z^n$$

The numbers *x*, *y*, *z*, and *n* are positive integers, and *n* is greater than 2. In contrast to the previous equations I have discussed, Fermat's Last Theorem states that this equation has *no* solutions.

In an era before scientific academies and scientific journals, when most universities did not even have a professor of mathematics, Mersenne was the focal point of mathematics in France. He held regular meetings at his convent and kept in touch with nearly every mathematician in Europe. If you wanted to publicize a new discovery, you would send it to Mersenne. The rest of the world would find out soon enough.

Fermat himself never visited Paris, never ventured out of the south of France, and met Mersenne only one time, in 1646. He adamantly refused to have anything published under his own name. Nevertheless, his results became known everywhere, thanks to Mersenne, and other mathematicians in France and abroad avidly desired to learn his methods.

Yet Fermat was very close-lipped. His normal *modus operandi* was to send his discoveries as problems to other mathematicians, often artfully concealed so that the true nature of his discovery would not be apparent to the recipient unless they had been working on similar problems themselves. In this way Fermat could ascertain whether he had found something new, without giving away what it was.

Of course this sort of challenge both tantalized and annoyed other mathematicians. René Descartes called Fermat a "braggart," and Bernard

Frénicle de Bessy accused him of posing impossible problems. Fermat seems to have been torn between his desire for recognition and a nearly pathological fear of revealing too many of his secrets. On the one hand, he was fond of quoting a motto of Sir Francis Bacon: *Multi pertranseant ut augeatur scientia* ("Many must pass by in order that knowledge may grow"). On the other hand, by his reluctance to publish, Fermat made many parts of his own work inaccessible to "passersby."

FERMAT WAS THE FIRST modern European mathematician to take an active interest in number theory, the study of equations with integer solutions. An ongoing theme in his work was Pythagorean triples: in other words, whole numbers a, b, and c such that $a^2 + b^2 = c^2$. As Fermat knew from studying a book that had been recently translated from the ancient Greek, Diophantus' *Arithmetica*, the Greeks had a general method for solving this equation. Fermat came up with innumerable variations on the theme: finding two Pythagorean triples with the same hypotenuse c; Pythagorean triples whose areas were square numbers, or twice a square, or such that the sum of the legs $a + b$ was square. He was able to resolve all of these problems to his satisfaction, even on occasion proving that there was no solution. (For example, no Pythagorean triangle has an area that is a perfect square.)

One day, probably between 1636 and 1640, he came up with another variation: Could a cube be written as a sum of two cubes? More generally, did the equation $x^n + y^n = z^n$ ever have whole-number solutions if the exponent n was greater than 2? In the margin of his personal copy of Diophantus, Fermat wrote: "No cube can be split into two cubes, nor any biquadrate [fourth power] into biquadrates, nor generally any power beyond the second into two of the same kind. For this I have discovered a truly wonderful proof, but the margin is too small to contain it." This handwritten note, which Fermat never intended anyone to see, became one of the most famous quotes in mathematical history. As number theorist André Weil has written, "How could he have guessed that he was writing for eternity?"

After Fermat died, his son Samuel collected and published his writings, including the copy of Diophantus with all of Fermat's marginal notes. In the 1700s, the Swiss mathematician Leonhard Euler took as a personal challenge to (re)-prove all of Fermat's results in number theory. The statement about splitting up powers into like powers was the only one that eluded him. He did prove that the equations $x^3 + y^3 = z^3$ and $x^4 + y^4 = z^4$ have no integer solutions, but he despaired of finding a general method for all n.

Fermat's innocent marginal note became known as "Fermat's Last Theorem." Technically, of course, it was not a theorem (i.e., a proven fact), but a conjecture. In 1825, Peter Gustav Lejeune Dirichlet proved that there are no whole-number solutions if $n = 5$. In 1839, Gabriel Lamé proved likewise for $n = 7$. By 1857, Ernst Kummer had proved it for all exponents n up to 100. Even though progress seemed agonizingly slow, the efforts to prove Fermat's Last Theorem were opening up new areas of mathematics, today called algebraic number theory.

In the twentieth century, Fermat's Last Theorem continued to spawn new mathematics, like the goose that laid golden eggs. In the early 1980s a German mathematician, Gerhard Frey, realized that any putative solution to Fermat's equation, $a^n + b^n = c^n$, could be used to construct an auxiliary curve, given by the equation $y^2 = x(x - a^n)(x + b^n)$, which struck Frey as a highly bizarre specimen. It was so bizarre, Frey argued, that it would violate another unproven conjecture in number theory, called the Taniyama–Shimura conjecture. The evidence was circumstantial at first, but subsequently an American mathematician, Kenneth Ribet, proved that Frey was right—if the Taniyama–Shimura conjecture was true, so was Fermat's Last Theorem.

Frey's idea was forehead-smackingly clever. He turned the variables in Fermat's equation into coefficients of a different equation. It's like the reversal of the foreground and background in a picture by M.C. Escher. Even so, it was very far from obvious that Frey's and Ribet's work represented any kind of breakthrough. They had only exchanged one seemingly unattainable goal for another. In effect, Frey and Ribet said: You want to climb Mount Everest? It's easy. Just grow wings.

Only one person in the world actually believed that he could prove the Taniyama–Shimura conjecture: Andrew Wiles. And he did it more or less by "growing wings." Actually, he built an airplane. Over a seven-year period, working alone in his attic, he linked together three of the most difficult, abstract, powerful theories of twentieth-century mathematics—the theories of L-functions, modular forms, and Galois representations—into a smoothly

functioning machine. One might compare his proof to the *Apollo* missions to the Moon, which combined (at least) three independent technologies: rocketry, computing, and communications. None of these technologies were developed with a Moon mission in mind. If any one of the three had been missing, the Moon missions would have been inconceivable. Yet they did come together, at just the right time, to conquer a famous "unsolved problem" (How can humans fly to the Moon?). Coincidentally, like Fermat's Last Theorem, that problem had been around for just about 350 years.

WILES ANNOUNCED HIS PROOF of Fermat's Last Theorem in 1993. Unlike Fermat, Wiles submitted his proof for publication, in 1994. In the three-and-a-half centuries between Fermat and Wiles, mathematicians had learned their lesson: a "theorem" without a published proof is no theorem at all. In fact, as Wiles wrote up his proof in 1993, he discovered a gap that took him a year (plus the assistance of a student, Richard Taylor) to plug. Perhaps, if Fermat had taken the trouble to write down his proof, he would have discovered a gap as well.

And this brings us to an inescapable question: Did Fermat actually find a correct proof? The answer of any competent number theorist would be a resounding no. According to André Weil, we can be certain that Fermat had a proof for the $n = 4$ case, and we may plausibly believe that he found something like Euler's proof for the $n = 3$ case. Both of these cases were solvable with Fermat's "technology." But beginning with $n = 5$, the problem changes very significantly. The case $n = 5$ required the nineteenth-century machinery of complex numbers and algebraic number fields. And, as I have described, Wiles' proof of the general case required top-of-the-line twentieth-century concepts that Fermat could never have dreamed of.

To the mathematical argument Weil adds a psychological one. Fermat repeatedly bragged about the $n = 3$ and $n = 4$ cases and posed them as challenges to other mathematicians (including poor Frénicle). But he never mentioned the general case, $n = 5$ and higher, in any of his letters. Why such restraint? Most likely, Weil argues, because Fermat had realized that his "truly wonderful proof" did not work in those cases. Every mathematician has had days like this. You think you have a great insight, but then you go out for a walk, or you come back to the problem the next day, and you realize that your great idea has a flaw. Sometimes you can go back and fix it. And sometimes you can't.

Weil's mathematical and psychological arguments are compelling. However, I would like to give the last word to a class of high-school students I taught in 1990, three years before Wiles announced his proof. On the last day of the course, a group of them performed a skit based on Fermat's life. As the curtain came down, they chanted in unison:

"Fermat! Fermat! He's our man! If he can't prove it, no one can!"

10

an unexplored continent
the fundamental theorem of calculus

William Dunham made an analogy between the discovery of Cardano's formula for the cubic and Columbus's discovery of America. However, the analogy falls short in one very important way. Columbus discovered an entire continent, comparable in size and importance to Europe. By contrast, Cardano's formula today is little more than a curiosity, even to mathematicians. Perhaps its significance could be compared to the impact of Columbus's discovery if Cuba and Hispaniola (where Columbus first made landfall) had merely been isolated islands with no continent nearby. It would certainly have been an amazing discovery, but perhaps not one to alter world history.

In the seventeenth century, though, mathematicians did find their equivalent of the New World, an unexplored "continent" of mathematics. The continent is called Calculus, and it had two primary discoverers: Isaac Newton and Gottfried Wilhelm Leibniz.

Calculus gave mathematicians and scientists a vocabulary for talking about quantities that change. The Fundamental Theorem provides a practical tool for solving problems about such quantities. Modern science, especially physics and engineering, would be inconceivable without it. Ironically, modern research mathematicians almost never use the term

$$\int_a^b f(t)\,dt = F(b) - F(a)$$

$$\frac{dF}{dt} = f(t)$$

The functions $f(x)$ and $F(x)$ are continuous functions of a variable x. In the second equation, t is an auxiliary variable. $F(x)$ is an antiderivative of $f(x)$, meaning that $dF/dx = f(x)$. The integral $\int$ and derivative d/dx were operations introduced by Newton, based on the ancient problems of finding tangents to curves and areas of regions with curved edges. The Fundamental Theorem says that these are inverse operations: if you integrate the derivative of any function, or vice versa, you get back the original function.

"calculus." The branch of mathematics that deals with functions, integrals, derivatives, and infinite series—in other words, everything connected with the Fundamental Theorem of Calculus—is called "analysis," and it is subdivided into Real Analysis, Complex Analysis, Functional Analysis, etc., just as America is subdivided into North, South, and Central America. To some extent the difference is one of intellectual rigor. In a "calculus" book, a mathematician will expect the arguments or explanations to be informal, intuitive, or entirely absent; in a book on "analysis," he or she will expect formal and correct proofs. However, in my opinion, the distinction is also motivated (or perpetuated) by intellectual snobbery.

Newton and Leibniz were born only four years apart—Newton in a village called Woolsthorpe, England, in 1642, and Leibniz in Leipzig, Germany, in 1646. Newton became a national hero in England, and was buried in 1727 in Westminster Abbey, the final resting place of kings. Leibniz, in spite of his successes in both mathematics and philosophy, was relatively unappreciated in his home country. When he died in 1716, he was buried in an unmarked grave.

Newton made fundamental advances in physics as well as mathematics: He invented the reflecting telescope and formulated Newton's laws of motion,

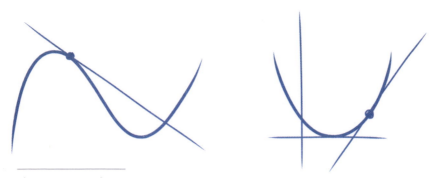

Above Tangents and curves.

which will be discussed at greater length in chapter 11. It is no exaggeration to say that our buildings stand and our spacecraft fly because of Newton's laws. His long and productive scientific career more or less coincided with his years at Cambridge University, where he arrived in 1661 as a subsizar (undergraduate), and left in 1696 to manage the British mint.

Leibniz, like Newton, had many interests outside mathematics. As a philosopher, he wrote (for example) about the problem of evil, and argued that although some evil was necessary, God had created the "best of all possible worlds." This belief was later ridiculed in Voltaire's famous book, *Candide*. Leibniz's mathematical work was concentrated mostly between the years of 1672 and 1676, when he was stationed in Paris as a diplomat and had plenty of time on his hands. It was perhaps the best place in the world to learn mathematics, because much of Mersenne's former network of friends was still intact, and had recently become formally organized as the French Academy of Sciences.

With the wisdom of hindsight, we can see that European mathematicians had been groping toward the discovery of calculus for the entire seventeenth century. Their attempts came from two separate directions. The first was the problem of quadrature, finding the areas of irregular (usually curved) regions. The problem of quadrature had fascinated mathematicians, of course, ever since antiquity. Early methods of computing areas were based on cut-and-rearrange arguments. Later, mathematicians like Archimedes in Greece and Liu Hui in China had become more sophisticated, approximating curved regions by a sequence of ever more accurate polygonal (straight-sided) regions.

In the early 1600s, an Italian mathematician, Buonaventura Cavalieri, had devised a systematic method, the "method of indivisibles," that involves cutting the unknown area up into narrow, rectangular slices and adding up their areas. In fact Archimedes had developed a similar method centuries before, but his work had been lost and was recovered only in 1906, too late to materially influence the history of European science. Neither Cavalieri nor Archimedes understood how to turn this method into a practical calculation tool; the examples they worked out were few and arduous.

THE SECOND ROUTE TO CALCULUS began with the problem of drawing tangents to arbitrary curves. That is, how can you draw a line that just grazes a curve at one point? Like the quadrature problem, the solution in general requires a sequence of approximations. In order find the tangent line at a point on a curve, you need to know the slope of the curve at that point. To compute the slope, you can imagine taking a nearby point on the curve, drawing a line segment between the nearby point and the given point, and computing the slope of that line segment. Your answer will always be slightly off the mark. If only you could make the nearby point "infinitely close" and the line segment "infinitesimally short"! Unfortunately this is difficult to justify mathematically, because it amounts to dividing 0 by 0.

Some people, including Fermat and probably also Newton's teacher at Cambridge, Isaac Barrow, had actually worked on both the problem of quadrature and the problem of tangents. However, only Newton and Leibniz grasped that the problems are actually flip sides of one another. If you have not studied calculus before, you should be shocked by this statement. There is absolutely no apparent connection between the tangent to a curve and the area inside a (different) curve.

The connection between the two ancient problems appears only after what seems at first like a highly arbitrary and artificial step: We turn curves into graphs. Of course, nowadays almost everybody is familiar with graphs. For example, there are graphs of stock prices in the business pages of the newspaper and electrocardiograms on monitors in hospitals. But in the seventeenth century, the idea of a graph was still very new.

A graph is a visual representation of the relationship between two variables: stock price and time, or electric potential and time. Some rule or process takes one variable (say, the time, or *t*) as input and produces the other variable (say, the stock price or *f(t)*) as output. The nature of this rule is not always clear in real-life examples like those that have been cited here. However, classical mathematicians were not interested in stock prices and electrocardiograms. They were interested in circles, parabolas, ellipses, spirals, and the like. For such curves, with a judicious choice of coordinate axes, it is often possible to write down a mathematical expression *f(t)* whose graph is the desired curve.

Below Color copper engraving of Leibniz (1646–1716), German philosopher and mathematician.

THE TWO ANCIENT PROBLEMS of tangency and quadrature are now more easily reinterpreted. The slope of a tangent line to a curve is actually a camouflaged version of the rate of change of the function it is a graph of. For example, suppose you went on a car trip, and at regular time intervals you recorded the distance on your odometer (let's call that *F(t)*), while at the same time recording the speed on your speedometer (let's call that *f(t)*). Thus, if you start at 12:00 and travel 60 miles in one hour, then *F*(1:00) = 60 miles. If you are going 30 mph at 1:00, then *f*(1:00) = 30 mph.

Now compare the odometer reading at time *t* to the next odometer reading, at time *t'*. These will give you two nearby points on the "odometer graph." To find the slope of the graph of *F(t)*, at time *t*, you would divide the distance you

E49.05 GOTTFRIED WILHELM von LEIBNEZ (1646-1716).
redit: The Granger Collection, New York

traveled in that short time interval by the elapsed time, in accordance with the high-school definition of slope as "rise over run." But that is the same thing as computing the average speed! (For example, if you went 2 miles in 3 minutes, your average speed over that time was $\frac{2}{3}$ miles per minute, or 40 mph.) Thus, the rate of change of the "odometer function"—over a short time interval—is the average of the speedometer function over that length of time.

There is just one more change to make in order to express the relationship in calculus terms. The words "over a short time interval" must be erased and replaced with the word "instantaneous." The instantaneous rate of change, or derivative, of the odometer function is the speedometer function. This seems like a tiny, almost insignificant modification. In fact, it is by far the hardest part of the whole argument. Neither Leibniz nor Newton fully justified it, and the debate over what exactly this word "instantaneous" means continued well into the nineteenth century.

However, let's give Newton and Leibniz the benefit of the doubt and move on to the second classical problem, the problem of quadrature, this time focusing on the "speedometer function," $f(t)$. Again, there are two times, a and b (this time they don't have to be close), and the aim is to discover the area (quadrature) of the region under the graph of $f(t)$, and between the times a and b. Skipping the explanation and going straight to the answer (you can read the explanation in any book on calculus, if you're brave), the quadrature of the "speedometer function" is the "odometer function." In calculus lingo, $F(t)$ is the *integral* of $f(t)$.

THUS, NEWTON AND LEIBNIZ introduced two new mathematical concepts: the derivative, which solves the tangency problem, and the integral, which solves the quadrature problem (although Newton used different words for these.) To some extent, both of these things had been done before; the integral was essentially the same as Cavalieri's method of indivisibles. But no one had realized before that the derivative and the interval are inverse operations. The derivative of the odometer function is the speedometer function; the integral of the speedometer function is the odometer function.

This inverse relationship is known today as the Fundamental Theorem of Calculus. Here is how we write it today as a formula (actually, it consists of two formulas):

$$\int_a^b f(t)dt = F(b) - F(a)$$

$$\text{and } \frac{dF}{dt} = f(t)$$

The first formula says that you can find the distance traveled, $F(b) - F(a)$, by integrating the speed. (That is what the symbol $\int$ means.) The second formula says that the speed is the rate of change, or derivative (that is what the symbol $\frac{d}{dt}$ means), of the distance. Thus either one of the functions $f(t)$ and $F(t)$ can be calculated if you know the other.

It is as if one sailor sailed west from Europe to find China, and another sailed east from China to find Europe, and they met in the middle and shook hands at Panama. The first equation (metaphorically speaking) says that sailing west from Europe gets you to Panama, and the second one says that sailing east from China gets you to the same place.

WHY DID THIS DISCOVERY open up a new continent of mathematics? It finally gave mathematicians absolute control over the concept of continuous change. Remember that continuous motion had puzzled the Greeks ever since Zeno put forth his famous paradoxes. Before Leibniz and Newton, mathematicians had been limited to static diagrams or discrete quantities. The world of continuous motion and continuously varying quantities was closed to them. But modern science is all about change. In calculus, mathematicians found the necessary vocabulary to do modern science.

Right Illustration from *The method of fluxions and infinite series* by Isaac Newton (1642–1727), first published in London in 1736.

Calculus is an immensely practical tool. Before Newton and Leibniz, finding an area or computing a slope was an unbelievably laborious process. But one of the benefits of

Τὰ κοινὰ κοινῶς, τὰ καινὰ κοινῶς.

Above Title page of "Acta Eruditorium Anno, 1684," Leipzig, 1684, in which appeared Gottfried Wilhelm von Leibniz's paper on his discover of the differential calculus, "Nova Methodus pro Maximus et Minimus

having two routes to Panama (to continue the analogy from above) is that you can pick the route that is more convenient. Very often, one route will turn out to be much more convenient.

With calculus, a mathematics student can now, in an afternoon, compute a better approximation of pi than Archimedes or Liu Hui. In fact, both Newton and Leibniz delighted in finding new expressions for pi and other constants. Tables of logarithms and sines—indispensible to mathematicians, engineers, and astronomers in the pre-computer age—could be computed routinely, and as precisely as patience allowed. Volumes, areas, and lengths of curves that had taken mathematicians centuries to figure out were now computable. Even as recently as the 1630s, Descartes had written that it was impossible to rectify a curve—that is, to find a straight line of equal length. With calculus, even a student can do it.†

Curiously, Newton was very reticent about discussing what he called the method of "fluxions." He apparently discovered the Fundamental Theorem between 1664 and 1666, but showed his work only in bits and pieces to a handful of people. Mathematicians had not yet realised that publishing, not hoarding secrets, was the surest route to progress.

---

† Here I have to confess that an undergraduate student can only succeed in a relatively small number of cases. However, beginning in the 1800s and continuing to the present day, the more difficult cases, such as rectifying an ellipse or a lemniscate, led to deep and beautiful new theories.

When Leibniz started thinking about rates of change and infinite sums in the 1670s, he surely had second-hand information about what Newton claimed he could do: namely, that he could compute infinite sums, areas, arc lengths, and so on. Sometime between 1673 and 1675, Leibniz unlocked the Fundamental Theorem, too. At this point he contacted Newton, to find out exactly what Newton knew, and proposed a sort of exchange of information: you tell me this, and I will tell you that.

Newton wrote back very cautiously, sending Leibniz only two letters. In the second, he divulged the Fundamental Theorem of Calculus—but concealed it in an indecipherable anagram. It is obvious that Newton did not want to share his discovery with Leibniz—he only wanted to be able to prove that he had gotten it first, in case Leibniz should later claim it as his own.

UNFORTUNATELY, THAT IS EXACTLY what then happened. Leibniz published his version of calculus in 1684 in a book called *Nova methodus*, while Newton, amazingly, waited until 1704 before publishing his first account of the method of fluxions. An extremely bitter dispute ensued over who should be known as the discoverer of calculus. English mathematicians supported Newton, while continental scholars mostly sided with Leibniz. Each side accused the other of plagiarism.

The consensus of modern historians is that they both were wrong, and they both were right. There was no plagiarism on either side, and both men independently made the same discovery. Newton undoubtedly was aware of the Fundamental Theorem first, but as I have said before, it does no good to discover America and then keep it to yourself. Leibniz was the first to tell the world about calculus. Partly for that reason, and partly because Leibniz's notation was simpler, the notation we use today is almost entirely due to Leibniz. No one today talks about "fluxions" and "fluents"—the words died with Newton.

11

of apples, legends … and comets
newton's laws

Ask most people what they know about Isaac Newton, and there is a good chance that they will tell you about an apple falling from a tree. According to legend, Newton was inspired to formulate his universal law of gravitational attraction by witnessing the fall of an apple, and realizing that the same force that explained its motion could also explain the motion of the planets. In some more recent embellishments, perhaps Newton was inspired by being hit on the head by the apple.

Here is an equally unverifiable counterlegend, which first appeared in print in 1858, in a delightful English journal called *Notes & Queries*. According to a contributor named "W.", Karl Friedrich Gauss—the leading mathematician of the day—dismissed the legend as follows: "The history of the apple is too absurd. Undoubtedly, the occurrence was something of this sort. There comes to Newton a stupid importunate man, who asks him how he hit upon his great discovery. When Newton had convinced himself what a noodle he had to do with, and wanted to get rid of the man, he told him that an apple fell on his nose; and this made the matter quite clear to the man, and he went away satisfied."

What is one to make of such legends? In reality, there is more substance to the apple story than Gauss realized (if the quote from him is authentic).

$$F = ma$$

$$F = \frac{G\,M\,m}{r^2}$$

The first equation is Newton's Second Law of Motion, the second is Newton's Law of Universal Gravitation. In both equations, F represents a force. The symbol *a* represents the acceleration of an object with mass *m*. In the law of universal gravitation, F is specifically the gravitational force between masses *m* and *M*, while *r* represents the distance between the objects. G is the universal gravitational constant, 6.672×10^{-8} cm^3 g^{-1} sec^{-2}.

The story is attested by two sources, one of whom was the famous French writer Voltaire, who heard it personally from Newton's niece. Hardly the sort of "stupid importunate man" that Gauss envisioned! One might think of the story as a highly encoded version of what actually happened. If you do not know the code, then you end up with the cartoonish story that Gauss so vigorously objected to.

Was there an apple tree? Yes. It was located at Isaac Newton's homestead in Woolsthorpe, England. He had lived there until 1661, when he went to Trinity College at Cambridge, and most importantly he returned there in 1665, when the last major outbreak of plague struck England. For close to two years Newton remained in his rural sanctuary. Those were the two years during which he developed the basics of calculus and began thinking about planetary motions. Newton wrote, "In those days I was in the prime of my age for invention and minded Mathematics and Philosophy more than at any time since."

However, Newton did not need the fall of an apple for inspiration. Gauss was right about that. Newton was surely inspired by the problem itself, which already had centuries of history behind it. Did the Moon, the Sun, and the planets require some sort of external agency to make them move?

If so, what was it? Aristotle had argued that heavenly bodies were made of different stuff than Earth, and that their natural motion was circular. Kepler thought that a propulsive force was needed to keep the planets in their orbits. Descartes more or less agreed; in his elaborate theory, the universe was composed of vortices that swept the planets along in their orbits. It is only natural that Newton, as a young scholar, would have been passionately interested in one of the leading scientific debates of the day. He worked out a system in which apples are subject to the same forces as planets. (This is why the apple is important! It refutes Aristotle.) And remarkably, the planets are in free fall at all times; they require no propulsion. (This refutes Descartes and Kepler.)

Opposite The first reflecting telescope, made by Issac Newton in 1668, stands by his manuscript of *Principia Mathematica*.

While the apple story has some merit, it fails to explain how Newton convinced the rest of the scientific world that his theory was correct. His masterpiece, *The Mathematical Principles of Natural Philosophy* (often called the *Principia* after its Latin title), set out to do for physics exactly what Euclid had done for geometry. At the very outset Newton stated three axioms: three laws of motion that all material objects obey, whether they be apples or moons. Later he added the law of universal gravitation, which quantifies how objects attract one another through gravity. From these principles alone, he proved that planets orbiting the Sun obey Kepler's three laws.

In fact, Kepler's first law is probably the main reason Newton wrote the *Principia*. Several other physicists—notably Newton's rival Robert Hooke, the architect Christopher Wren, and the Dutch physicist Christian Huygens—had also arrived at "Newton's" law of universal gravitation by the early 1680s. But they had been unable to show that the law causes planets to orbit in ellipses; they could only account for the mathematically much simpler case of circular orbits. In 1684, Newton's friend Edmund Halley asked Newton if he could prove that planets had elliptical orbits. Newton said that he could, and Halley cajoled him into putting his argument into print. The result, three years later, was much more than the solution of one problem; it was the blueprint for all future physics books.

Halley, who paid for part of the printing costs out of his own pocket, was eventually rewarded for his efforts in a very unique way. Newton's theory

applies to comets, as well as to apples and planets. (In fact, Newton himself emphasized this point.) Because comets follow elliptical orbits, they must return over and over again. Halley realized that one comet in particular had been seen repeatedly at roughly 75-year-intervals: in 1456, 1531, 1606, and 1682. Thus he predicted, correctly, that it would return in 1758 (long after his own death). It has continued to return every 75 to 76 years ever since then, and is now known as Halley's Comet.

NEWTON'S FIRST LAW states that a moving object will continue moving in a straight line forever, unless some external force stops it or changes its path. This seems quite surprising at first: after all, golf balls don't keep going forever, and planets don't move in straight lines. In both cases, the reason is that there are external forces acting on the object. In the case of the golf ball, the forces are gravity, wind resistance (while the ball is in the air), and friction with the ground after it lands. In the case of planets, the hidden force is the Sun's gravity.

Newton's second law says that the force on an object equals the rate of change of its momentum. In the language of calculus, we would say that:

$$F = \frac{d}{dt}(mv)$$

recalling that d/dt denotes the rate of change and mv (where m is the mass of the object and v is its velocity) denotes the momentum. In most applications the mass of the object does not change, and in this case Newton's second law becomes $F = ma$ (i.e., force equals mass times acceleration), a formula that is today memorized by every beginning physics student.

Newton's third law, "for every action there is an equal and opposite reaction," is somewhat less often used by physicists than the first two, but it explains, for example, why a rocket works. The action of propelling exhaust out of the rocket's nozzles creates a reaction: the acceleration of the rocket in the opposite direction.

Collectively, these three laws explain how all forces affect the motion of all solid bodies. On the other hand, Newton's law of gravitation pertains to one force only, the force of gravity. It states that the gravitational attraction between any two objects, one of mass M and the other of mass m, is:

$$F = \frac{-GMm}{r^2}\hat{\mathbf{r}}$$

The denominator r^2 indicates that the strength of the gravitational force is inversely proportional to the square of the planet's distance (r) from the Sun. (This is the part of the formula that Hooke, Wren, and Huygens had already guessed.) The minus sign and the vector $\hat{\mathbf{r}}$ (read as "r-hat") indicates that the direction of the force is toward the Sun. In other words, Kepler and Descartes were wrong. There is no force pushing the planets forward in their orbits, only a gravitational force pulling them to the side (that is, toward the Sun).

Newton's truly novel accomplishment was his ability, using calculus,[‡] to combine the law of gravitation with his laws of motion to set up—and then solve—equations describing a planet's orbit. Together, his physical insight and his mathematical tools ushered in a new era of celestial dynamics, when the motion of planets—and eventually, rockets and spacecraft—could be predicted and controlled, rather than merely observed.

[‡] It is often claimed that Newton deliberately avoided the use of calculus in the *Principia*, rewriting all the proofs in terms of Euclidean geometry. It is true that he avoids the *notation* of calculus, but his work is fully imbued with the *ideas* of calculus.

12
the great explorer
euler's theorems

In 1988, the magazine *Mathematical Intelligencer* organized a poll to determine the most beautiful mathematical theorems in history. Amazingly, four of the top five theorems on the list were proved by the same man: Leonhard Euler. Even more remarkably, it is easy to come up with a half-dozen more theorems by Euler that could have made the list. In fact, Euler authored more than 800 articles and about 50 books and memoirs. The Academy of Sciences in St. Petersburg (where he spent the last 17 years of his life) was unable to keep up with his output, and continued publishing articles by Euler for half a century after his death!

Leonhard Euler was born in Basel, Switzerland, in 1707, and remained proud of his native country and town throughout his life, even though he never again set foot in Basel after age twenty. He had the good fortune to come of age in an era when mathematics was beginning to turn from a scholarly pursuit into a profession. England's Royal Society had been founded in 1660, and the French Academy of Sciences soon afterward, in 1666. Gottfried Wilhelm Leibniz, after returning from France, persuaded King Frederick I to establish the Prussian Academy of Sciences in 1700. By the early 1720s, when Tsar Peter I of Russia was building his new capital at St. Petersburg, an academy of science was almost *de rigueur* for a royal

$$e^{ix} = \cos(x) + i\sin(x)$$

The number $e = 2.718281828459045...$ is the base of the natural logarithm function and the second most ubiquitous constant in mathematics, after π. The letter i represents the imaginary unit, $i = \sqrt{-1}$. The functions cos, and sin are the cosine function, and sine function respectively.

court. Monarchs had begun to realize that mathematicians and scientists could play an important role in building their countries' infrastructure and military prowess.

In 1724, Peter I founded the Russian Academy of Sciences and invited a number of foreign scientists to move to his raw new capital. At that time, the opportunities for a mathematician in Switzerland were limited, so Euler seized the opportunity. During his first period in Russia, from 1727 to 1741, Euler's reputation throughout Europe rose rapidly. In 1735, he stunned the world of mathematics by evaluating an apparently simple infinite sum, that no one had been able to crack:

$$1 + \frac{1}{2^2} + \frac{1}{3^2} + \frac{1}{4^2} + ...$$

demonstrating that it is equal to:

$$\frac{\pi^2}{6}.$$

Euler's argument is explained in masterful fashion in George Pólya's 1954 book *Mathematics and Plausible Reasoning, Part I*. For readers who know their trigonometry and have some knowledge of infinite series, it is one of the best expositions of how a mathematical genius thinks (both Pólya and Euler).

Euler continued to move from triumph to triumph. His book *Mechanica*, published in 1736, took mechanics out of the realm of Euclidean geometry, where Newton had awkwardly placed it, and rephrased it in the much more appropriate language of calculus. In 1738, Euler won the Grand Prix de Paris competition for the first time.[*] During this period, Euler introduced notation that is used by all mathematicians today: e for the base of the natural logarithms, i for $\sqrt{-1}$ and $f(x)$ for functions.

Unfortunately, after Empress Anna died in 1740, a backlash set in against the foreigners whom Peter had invited to "cut a window through to Europe." Euler found his situation increasingly untenable, and accepted an invitation from Emperor Frederick II of Prussia to join the academy of sciences in Berlin.

If Euler in St. Petersburg was an up-and-coming star, Euler in Berlin was a mature scientist at the height of his power. Among other things, during this period he resuscitated Fermat's number theory and re-proved most of the things that Fermat had claimed to prove (except, of course, the Last Theorem). He worked out how Newton's laws of motion apply to fluids, deriving what are still known as Euler's equations of hydrodynamics. He published books on topics ranging from calculus to naval science. And he worked on some of the ongoing puzzles of astrophysics.

Although Euler prospered in Berlin, the one person he never managed to impress was his employer. The mercurial monarch, Frederick II, was greatly attracted to pomp and culture, particularly French culture, and he could not abide the stolid Swiss mathematician in his court. Though he conceded that Euler was "useful," Frederick compared him to a Doric column, "anything but elegant." On another occasion, he wrote to Voltaire, "We have here a great Cyclops of mathematics," referring unkindly to the fact that Euler

---

[*] For well over a century, the most prestigious honors in mathematics were international competitions, organized by the national academies of science, in which papers were solicited on a particular topic. Euler won his first Grand Prix de Paris in 1738 for a paper on the nature of fire, and subsequently won eleven more (a record, of course).

LEONARD EULER.

London, Published as the Act directs, Oct.r 13.th 1804, by J. Wilkes

MITHODUS
INVENIENDI
LINEAS CURVAS
Maximi Minimive proprietate gaudentes,
SIVE
SOLUTIO
PROBLEMATIS ISOPERIMETRICI
LATISSIMO SENSU ACCEPTL
AUCTORE
LEONHARDO EULERO,
Profeſſore Regio, & Academiæ Imperialis Scientia-
rum PETROPOLITANÆ *Socio.*

SUPRA INVIDIAM

LAUSANNÆ & GENEVÆ,
Apud MARCUM-MICHAELEM BOUSQUET & Socios.
MDCCXLIV.

had lost the vision in his right eye. Euler longed to become president of the Berlin Academy of Sciences, but it was clear that under Frederick II this would never happen, so in 1766 he accepted Empress Catherine II's invitation to return to Russia.

During the last years of his life Euler's work continued unabated, even though a failed cataract operation in 1771 left him nearly blind. He gradually withdrew from the St. Petersburg Academy because of internal politics. However, in 1783, the last year of Euler's life, Princess Ekaterina Dashkova took over the directorship of the Academy and insisted on making her entrance with him. When she realized that the seat next to hers was taken, she wrote, "I therefore turned to Mr. Euler and told him to sit down where he thought fit, for any place he occupied would always be the first." Euler had finally found a place where he was appreciated.

TWO HUNDRED YEARS LATER, mathematicians still appreciate him. Let us take a look at the four theorems of Euler that the readers of *Mathematical Intelligencer* rated in the top five of all time:

$e^{i\pi} + 1 = 0$ (or $e^{i\pi} = -1$) Surely one of the most paradoxical statements in mathematics, it is often written in the former way because this allegedly "unifies" the five most important constants of mathematics: 0, 1, π, e, and i.

Here is what the "most beautiful equation in history" really means. Probably the most important function in calculus is the exponential function $\exp(x)$, because it is the only function that is its own derivative and its own integral. The name is apt because the values of the function are all powers of the number exp(1). For instance, exp(2) = (exp 1)^2, exp(3) = (exp 1)^3, and so forth. To save space, we can call the number exp(1) "e", as Euler did. Then the seemingly nonsensical number e^π (e multiplied by itself π times, whatever that means) can be defined as exp(π). Because the function exp(x) is defined by calculus it can be computed using calculus. Thus we can determine that exp(π) = 23.1406…

But what does it mean to raise a number to an imaginary power? How do we multiply 23, or indeed any other number, by itself $\sqrt{-1}$ times? Surely this is mathematics run amok.

Again, the trick is not to think of numbers but functions. Euler knew a way to write the function exp(x) as an infinite sum. With this equation, it was a simple matter for Euler to substitute ix in place of x, keeping in mind that $i^2 = -1$, $i^3 = -i$, $i^4 = 1$, and so forth. The result is:

$$\exp(\,ix\,) = 1 + ix - \frac{1}{2}x^2 - \frac{1}{6}ix^3 + \frac{1}{24}x^4 + \ldots$$

Separating the terms without i from the terms with i, Euler instantly recognized the second and third most important functions of calculus, the sine and cosine functions:

$$\exp(ix) = \cos(x) + i\sin(x)$$

This is the formula that Euler himself considered important! It is featured in his calculus textbook of 1748. Nowhere in that textbook, or anywhere else, did he write the equation that has become associated with his name ($e^{i\pi}$ = −1). Euler understood that calculus was about functions, not about numbers. However, we can get the "number version" of his formula easily enough by the final step of substituting $x = \pi$. Then $e^{i\pi} = \exp(i\pi) = \cos(\pi) + i\sin(\pi) = -1 + 0i = -1$.

Beauty is, of course, in the eye of the beholder. The readers of *Mathematical Intelligencer* preferred the numerical formula because it relates the five fundamental constants of mathematics. One could argue that the version with exp, cos, and sin is much more beautiful, because it relates the three most fundamental functions of calculus, functions that have likewise been studied for centuries. Furthermore, it explains the meaning of the otherwise opaque equation, $e^{i\pi} = -1$. Surely a formula that helps us understand mathematics better is much more beautiful than a formula that only mystifies us.

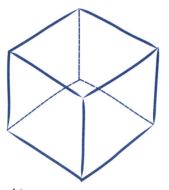

Above A cube has 8 vertices, 12 edges, and 6 faces.

$V - E + F = 2$. Second in the *Intelligencer* poll was this elegant formula, which relates the number of vertices (V), edges (E), and faces (F) of any polyhedron. For example, a cube has 8 vertices, 12 edges, and 6 faces; and, sure enough, $8 - 12 + 6 = 2$. As it turns out, this equation has exceptions that Euler was not aware of. For a doughnut-shaped polyhedron, for instance, $V - E + F = 0$, not 2. With hindsight, it is clear that this equation marked the beginning of a new branch of mathematics called topology, which flourished in the twentieth century. The number $V - E + F$ is now called the Euler characteristic. It is a "topological invariant" that distinguishes one two-dimensional surface from another. Sphere-shaped surfaces always have Euler characteristic 2; doughnut-shaped surfaces always have Euler characteristic 0; pretzel-shaped surfaces have Euler characteristic −4, and so on.

The infinitude of prime numbers. This was an ancient discovery, known to Euclid, but Euler discovered a radically different proof, which not surprisingly uses the concepts of functions and infinite series that were so dear to him. The proof involves the zeta function, $\varsigma(x) = 1 + 1/2^x + 1/3^x + 1/4^x + \ldots$ Euler showed that this infinite sum is also equal to Euler's product:

$$\zeta(x) = \cfrac{1}{\left(1 - \dfrac{1}{2^x}\right)\left(1 - \dfrac{1}{3^x}\right)\left(1 - \dfrac{1}{5^x}\right)\left(1 - \dfrac{1}{7^x}\right)\cdots}$$

The product in the denominator runs over all prime numbers (2, 3, 5, 7, …). For number theorists, Euler's product is probably the most important formula ever discovered. Most of what we know about the distribution of prime numbers comes from the careful study of the zeta-function: a theme that will be returned to later.

The Basel Problem. Finally, the fourth of Euler's equations, which was fifth place in the list, was the formula, already mentioned, that cemented his reputation:

$$1 + \frac{1}{4} + \frac{1}{9} + \frac{1}{16} + \ldots = \frac{\pi^2}{6}$$

The discerning reader will notice that the left-hand side is actually $\varsigma(2)$, and might even wonder if Euler's product for the zeta function has something to do with this formula. To the best of my knowledge, the answer is no. Euler actually derived it from an infinite product representation for the sine function, rather than the zeta function.

THE INTELLIGENCER LIST was, possibly, somewhat biased toward pure mathematics. There aren't very many formulas on this list that are used in non-mathematical applications. And that's a pity, because Euler could do it all. He developed the first theory of hydrodynamics; he studied the buckling of elastic rods; he even worked on the optimal placement of masts in ships (a very important practical problem of the day). It is doubtful that he saw much of a distinction between mathematics that was done for its own beauty and mathematics that was done to solve a practical problem.

Finally, as math historian Jeremy Gray points out, one of Euler's most important contributions to math was not an equation at all. Part Two has described throughout how controversies arose and progress slowed because mathematicians, for one reason or another, were reluctant to share their secrets. It is the one thing that del Ferro, Tartaglia, Galileo, Fermat, and Newton all had in common. Euler was the one shining exception. He published abundantly; he was willing to step aside and give others credit; his articles routinely delivered *more* than they promised. He led by example, and helped transform mathematics into what it is today—a profession where information is not proprietary but is (with some unusual and unfortunate exceptions) openly shared.

equations in a promethean age

If you ever go to Dublin,

Ireland, take a bus to Broombridge Road and get off at the Royal Canal. You may not realize it, but you have just arrived at the site of the most famous mathematical graffiti in history.

From street level, the stone bridge that the road takes its name from is small and nondescript, but if you descend to canal level and walk to the west side of the bridge, you will find (along with lots of modern, spray-painted graffiti) a plaque with the following inscription:

"Here, as he walked by on the 16th of October 1843, Sir William Rowan Hamilton in a flash of genius discovered the fundamental formula for quaternion multiplication $i^2 = j^2 = k^2 = ijk = -1$ & cut it on the stone of this bridge."

To be honest, no one knows if Hamilton really did carve his formula into Brougham (pronounced "broom") Bridge. The source of the story is a letter that he wrote to his son, Archibald, many years later, and like many family stories it may have been embellished. However, there is no doubt of Hamilton's excitement over the discovery of quaternions, which he considered the greatest of his life.

Hamilton became so besotted with his creation that he spent the rest of his life studying the equations. Viewed from more than a century later, the discovery was in fact a turning point in mathematical history, but in a subtler way than Hamilton could have anticipated. They were the first example of a new algebra, created entirely out of one person's imagination. This step, along with the nearly simultaneous discovery by other mathematicians of new geometries and new functions, liberated mathematicians from traditional structures (and strictures). For the first time, they could venture beyond the real world—they were free to invent entire new worlds.

Before the nineteenth century, there was only one algebra and one geometry. The idea did not even occur to mathematicians to invent anything different. It is true that the concept of "number" had gradually expanded over the centuries, first to include irrationals, then zero and negatives, and finally imaginary numbers. But these new kinds of number were annexed only with great difficulty, and only after bitter debate. They were accepted only because they were indispensible. Similarly, calculus was a revolutionary technique but it did not involve the creation of a new geometry. Newton's concept of space was exactly the same as Euclid's.

All of this changed in the nineteenth century. It was a revolutionary era, in mathematics as in the outside world. Beginning with the French Revolution, European societies were scrapping their old political structures and creating new ones. Likewise, mathematicians began trying out new structures that directly contradicted axioms they had been using for centuries. It was an era when Mary Shelley could write her novel, *Frankenstein; or the Modern Prometheus*, warning of the dangers of scientists playing God. Mathematicians became modern-day Prometheans, like Dr. Frankenstein, although their creations were not made of flesh and blood.

Hamilton would have deplored this development. Socially he was conservative, a supporter of the English Crown in an Ireland that was starving and chafing under English domination. Mathematically, he invented quaternions in order to understand Euclidean space, not in order to create a new algebra. Nevertheless, revolutions are often begun by people who have no inkling of what they are starting.

13

the new algebra
hamilton and quaternions

Born in 1806, William Rowan Hamilton was a child prodigy who knew all the European languages, as well as Hebrew, Latin, Greek, and others, by the time he was ten years old. Hamilton was an enthusiastic amateur poet throughout his life, and was a close friend of the English poet William Wordsworth. It was Wordsworth who delicately, but wisely, advised Hamilton that he had more to offer the world as a scientist than as a poet.

In 1827, Hamilton was appointed Royal Astronomer of Ireland—even though he had not yet graduated from university! The appointment had more to do with his research on optics than his interest in astronomy. He had already published papers on optics as an undergraduate, and five years later he made a sensation with his discovery of conical refraction. Certain kinds of crystals, which are called birefringent, can split a light beam up into two separate beams. Hamilton proved mathematically that if the angle of incidence was just right, the beam would split up not just into two beams, but into a hollow cone of light. Later that year Humphrey Lloyd demonstrated conical refraction in his laboratory. It was one of the first times that a new physical phenomenon had been deduced by pure mathematics first, and confirmed by experiment second. After this breakthrough, Hamilton was no longer just a prodigy; he was one of Great Britain's scientific heroes.

$$i^2 = j^2 = k^2 = ijk = -1$$

i, j, k represent imaginary units. Multiples of these units can be added to real numbers to form quaternions, *a + bi + cj + dk*. The above multiplication rules will then uniquely define the product (and, with a little work, the quotient) of any two quaternions.

Hamilton's other great discovery, quaternions, had a much longer and stranger history. Sometime around 1830 Hamilton began looking for a way of multiplying number triplets together. By this time the multiplication and division of number pairs, or complex numbers, had proven itself to be not only possible but an essential part of mathematics. Any two such pairs, say (*a* + *bi*) and (*c* + *di*), can be multiplied or divided, using the rules of algebra plus the miraculous identity $i^2 = -1$:

$$(a+bi)(c+di) = ac + bci + adi + bdi^2 = (ac - bd) + (bc + ad)i.$$
$$\frac{1}{(a+bi)} = \frac{(a-bi)}{(a^2+b^2)}.$$

But there was another motivation for multiplying number triples. Hamilton knew that complex multiplication has a geometric meaning, quite apart from its origins in algebra. For example, the algebraic operation "multiply by *i*" is the same as the geometric operation "rotate by 90 degrees counterclockwise." More generally, the instruction "multiply by (*a* + *bi*)" can be broken down into two steps—a rotation and a dilation. This interpretation takes a lot of the mystery out of complex numbers. Some people may have trouble imagining a

number whose square is −1, but everybody knows that two 90-degree rotations give a 180-degree rotation. Not only that, this description makes the invertibility of complex multiplication completely obvious. To undo the operation "rotate counterclockwise by 72 degrees," you simply rotate clockwise the same amount. To undo the operation "enlarge to 150 percent," you reduce to 67 percent.

Alas, complex numbers are limited to representing operations in a plane. They are great for manipulating two-dimensional photographs, but not three-dimensional reality. Hamilton was convinced that there must be an algebra of three dimensions as powerful as the two-dimensional algebra of complex numbers. But until 1843, he was stymied. The rock on which all his attempts foundered was division. No matter how he defined the multiplication of number triples, he was not able to divide them.

And then came Hamilton's Monday stroll across Brougham Bridge. What he suddenly realized—although his calculations must have subconsciously been leading him to this point—was that introducing a fourth number makes both multiplication and division possible.

Thus Hamilton proclaimed: Let there be three imaginary units, i, j, and k. Let them go forth and multiply using the following rules:

$$i^2 = j^2 = k^2 = -1$$

$$ij = -ji = k, jk = -kj = i, ki = -ik = j$$

Then one can add, subtract, multiply, and divide any two quaternions, $(a + bi + cj + dk)$ and $(w + xi + yj + zk)$, just by following the normal rules of algebra. Division is the trickiest part, of course. It turns out that $1/(a + bi + cj + dk)$ equals $(a - bi - cj - dk)/(a^2 + b^2 + c^2 + d^2)$, as you can check by multiplying both sides by

$(a + bi + cj + dk)$. Don't even think about what the imaginary numbers i, j, and k mean—just do it and you'll see that it works.

SUCH A WANTONLY Promethean act had almost never been seen before in mathematics. Hamilton's colleagues were aghast, though they could not find anything wrong with it. "There is still something in the system which gravels me," wrote his friend John Graves. "I have not yet any clear view as to the extent to which we are at liberty arbitrarily to create imaginaries, and to endow them with supernatural properties."

Hamilton spent the remaining 22 years of his life proselytizing the importance of quaternions. He wrote a 700-page book about them but then, convinced it was too difficult, started a shorter "manual" for students—which grew to more than 800 pages and lay unfinished when he died.

However, quaternions fell from popularity for a variety of reasons. First, Hamilton had intended to find an algebra of three-dimensional space. What, then, did the fourth dimension of a quaternion mean? Hamilton argued that it could represent time[*]—and in so doing, he became the first scientist to merge time and space into a single "spacetime." However, physics had not yet matured to the point where it needed this concept; it would have to wait for the twentieth century and Albert Einstein.

A second blow to quaternions was the development of vector analysis in the 1870s, by Oliver Heaviside (an Englishman) and Josiah Willard Gibbs (an American). Heaviside and Gibbs dispensed with the imaginary quantities entirely, and simply represented points in space by a triple of numbers, (a, b, c), called a vector. Instead of one multiplication, they defined two different vector multiplications, the dot product and the cross product. Neither one of them is invertible, and in fact the dot product of two vectors isn't even a vector—it's a real number. However, what they lack in elegance, vectors make up for in practicality. They are well adapted to the problems of physics and engineering. After a series of polemics in the 1890s between the

---

[*] It is a splendid irony that in Hamilton's quaternions, the three dimensions of space—the i, j, and k dimensions—are imaginary while time is the real coordinate (i.e., the number a in the expression $a + bi + cj + dk$). The real world and imaginary world have switched places!

quaternionists and the vector analysts, the analysts won. A typical example is this scathing comment from William Thomson, Lord Kelvin, in 1892: "Quaternions came from Hamilton after his really good work had been done; and though beautifully ingenious, have been an unmixed evil to those who have touched them in any way, including Clerk Maxwell." Nowadays, you will never see a quaternion in a freshman physics book.

One reason that vectors won is that neither Hamilton nor his followers really understood what quaternions were, and therefore they were trying to use them the wrong way. Just as complex numbers represent geometric operations (rotations and dilations in a plane), quaternions represent rotations and dilations of space. Thus, they are not vectors. A vector is something that is acted upon by rotations. Quaternions are the rotation itself.

HAMILTON NEVER FIGURED OUT the difference. That was left to a twentieth-century mathematician, Elie Cartan, who named this kind of quantity a spinor (discussed in Part Four) to distinguish it from vectors. His findings vindicate Hamilton's belief in quaternions, even though he did not grasp their significance. The key point to bear in mind is that quaternions are the very best way to represent anything that spins in three dimensions. That includes protons, neutrons, and electrons—the building blocks of our physical world. Of course, the existence of these subatomic particles was not even suspected in Hamilton's time. He had discovered the right mathematics almost a century before it would be needed.

But one immediate effect of quaternions, as mentioned above, was to liberate mathematicians to think about other kinds of algebra. Quaternion multiplication violates one previously unquestioned rule of algebra. It is not commutative; that is, the product of two quaternions is sensitive to the order in which they are multiplied. For example ij is equal to k, but ji is equal to $-k$. This was the first known example of a non-commutative algebra.

Hamilton's friend Graves soon got over his uneasiness and discovered an algebra of 8-tuples, or octonions. These are even more finicky than quaternions, because products of three octonions are sensitive not only to the order of the octonions but also to their grouping. If a, b, and c are real numbers, or complex numbers, or even quaternions, then (ab)

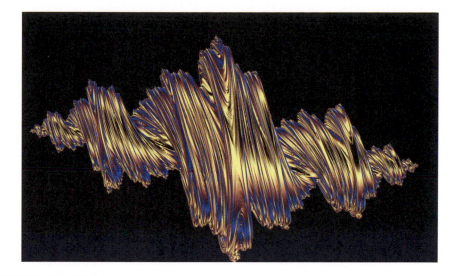

$c = a(bc)$, but for octonions, $(ab)c$ is usually *not* equal to $a(bc)$. The independence of grouping was a property that mathematicians had always assumed without even realizing it; Hamilton had to make up a new word for it—the associative law. It might seem as if the next step would be an algebra of 16-tuples. However, each doubling of the number of dimensions comes with a sacrifice. Going from 2 dimensions to 4, you lose commutativity. Going from 4 dimensions to 8, you lose associativity. And going from 8 dimensions to 16, you lose division. At this point, Hamilton's program of defining hypercomplex numbers breaks down, because the one thing he always insisted on was division.

Other mathematicians had no such compunctions. Once the floodgates were opened, anything was possible. You can have algebraic structures with three operations (addition, subtraction, multiplication) called rings; or with two operations (addition and subtraction, or multiplication and division), these are known as groups; or you can even pare it down to one operation; these structures are called monoids. With such a variety of algebraic structures to choose from, the question becomes not what is *possible*, but what is *worth studying*. Does a new structure help solve pre-existing problems? Does it have a deep, challenging, inherently beautiful theory? One new algebraic structure that has consistently scored high on both criteria is the concept of a group—and that is what will be discussed next.

14
two shooting stars
group theory

In the early nineteenth century, mathematics lost two of its brightest talents at a very young age, a 26-year-old Norwegian and a 20-year-old Frenchman. Niels Henrik Abel and Évariste Galois were linked by more, though, than their untimely deaths. They combined to give a definitive answer to one of the most classical questions in mathematics: Is there a universal version of Cardano's formula for the cubic (which I discussed in Part Two, page 61)? In the process, they opened up a new branch of mathematics, which we now call group theory.

Abel was born in 1802, the son of a long line of country clergymen. His childhood was a complicated time politically for Norway, which was a sort of pawn of the Napoleonic Wars. After almost 300 years of relatively benign Danish rule, Norway briefly became an independent state in 1814, but later that same year its parliament voted to recognize the Swedish king. Abel's father, twice elected to parliament, became a lightning rod for scandal because of his minority pro-independence views. After Abel's father died in 1820, his alcoholic mother left with another man, and Niels and his siblings were left in poverty.

Fortunately, Abel's teachers recognized and encouraged his talent for mathematics. By the time he finished his university studies, it was clear to

$$\text{Gal}\left(K/Q\right) = S_5$$

Gal(K/$\mathbf{Q}$) represents the Galois group of a polynomial over the rational numbers $\mathbf{Q}$. S_5 represents the group of all 120 possible permutations of five objects. Whenever a polynomial has a Galois group equal to S_5, the polynomial cannot be solved using the five basic operations: $+$, $-$, $\times$, $\div$, and n-th roots.

them that his abilities were much beyond any position that could be found for him in Norway. The faculty persuaded the Swedish king to give Abel a two-year travel stipend to visit Europe's leading centers of mathematics: Göttingen (in Germany) and Paris. Abel's extended *Wanderjahr*, from 1825 to 1827, started very auspiciously. One of the first people he met in Germany was Leopold August Crelle, who was about to start a new journal called *Journal of Pure and Applied Mathematics*, often simply known as *Crelle's Journal*. However, Abel failed to impress the other leading mathematicians of the day, such as Karl Friedrich Gauss in Germany and Augustin-Louis Cauchy and Adrien-Marie Legendre in France. During his visit to Paris in 1826, Abel submitted what he considered his most important paper to the Parisian Academy of Sciences. Cauchy apparently lost it in a desk drawer, and it was not printed until 1841, long after Abel's death.

By the time Abel got back home to Norway his other articles had started appearing, in rapid-fire succession, in *Crelle's Journal*. The Parisian mathematicians were astounded, first to read a series of breakthrough papers by an unknown mathematician from the hinterlands; then to learn that this mathematician had actually been in Paris; and finally, to learn that he had tried to present them a paper and they had lost it! Legendre sent his apologies to Abel, and along with three other mathematicians petitioned the Swedish king to find some way to assist "a young Monsieur Abel, whose works show he has mental powers of the highest rank, and who nevertheless grows ill there in Christiania [Oslo] in a position of too little value for one of his rare and early-developed talent."

Unfortunately, the rumor that Legendre and the others had heard about Abel's ill health was true. By 1828, Abel had developed tuberculosis, and in April of 1829 he died, just two days before a letter arrived from Crelle saying that he had arranged a professorship for Abel in Berlin.

ÉVARISTE GALOIS' STORY is also one of unbelievably bad luck, compounded by poor judgment. Born in 1811 near Paris, he was apparently a very difficult student in high school, described by his teachers as "original" and "bizarre." At the age of 17, he sent a paper on the solvability of polynomials to Cauchy—the same Cauchy who had lost Abel's manuscript a couple years earlier. Galois' paper was lost, too, though this time it was not Cauchy's fault. "I cannot in truth conceive of such carelessness on the part of those who already have the death of Abel on their consciences," Galois later wrote. The accusation is, of course, completely unfair. Though the loss of Abel's paper was a scandal, the Academy was in no way to blame for Abel's death and, as noted above, had even tried to intervene on his behalf. However, this invective does give us an insight into Galois' character. He was a rebel against authority, and the Academy became for him a symbol of tyrannical power.

In 1829, Galois joined a revolutionary organization called the Society of the Friends of the People. The following year, rioting broke out in the streets of Paris, and King Charles X was forced to abdicate the throne. Extreme Republicans (such as Galois) wanted to abolish the monarchy altogether, but moderate Republicans led by the popular Marquis de Lafayette prevailed. They named Louis-Philippe as the "citizen king" of France, a king who would be bound by constitutional restrictions.

Galois was unable to participate in the July 1830 revolution because he was a student at the École Normale, and the school's director literally locked the students in. However, by 1831 he had graduated, and he no longer had to sit on the political sidelines. He was arrested twice that year, once for threatening the life of King Louis-Philippe and the second time for participating (heavily armed) in a demonstration on Bastille Day. While he was in prison, he received news that the French Academy had rejected his latest paper on the theory of equations.

Galois was released from prison in April 1832, and by the end of May he was dead. The events that led to this outcome are far from clear. One historian has written a book arguing that it was a police-organized provocation, while others deny it. Galois himself wrote to his friends, in moving words, that he was forced to take part in a duel over a woman:

"I beg patriots, my friends, not to reproach me for dying otherwise than for my country. I die the victim of an infamous coquette and her two dupes. It is a miserable piece of slander that I end my life … I would like to have given my life for the public good. Forgive those who kill me for they are of good faith."

On May 30, the day after Galois wrote this letter, he was shot in the stomach by a man whom the writer Alexandre Dumas identified as Pescheux d'Herbinville, a hero of the Republican cause. Galois' opponent left him on the ground to die. He was found several hours later, still alive, but he died the following day.

ALTHOUGH IT IS TEMPTING to wonder what Abel and Galois might have achieved if they had both lived, in fact they both accomplished a great deal in their short lives. Their lasting fame rests on the theorems they proved, and not on the way that they died.

Both Abel and Galois were fascinated by the problem of finding the solutions of polynomial equations. In Part Two, I recounted how Cardano "stole" the secret for solving cubic (third-degree) polynomials, and how his servant Ferrari subsequently discovered a method for solving quartics (fourth-degree). In both of these formulas, the solutions, or "roots," can be expressed using only the operations of algebra ($+$, $-$ $\times$, $\div$, and nth roots for any n). Often the nth roots, or "radicals," are nested inside one another, a square root inside a cube root inside a fourth root; this accounts for the term "solution by radicals." However, no one in the intervening three centuries had found a universal solution by radicals for fifth or higher degree equations, and some were starting to suspect that no solution could be found.

It is difficult indeed to prove that a task is impossible. It is not just a matter of trying and failing to solve it. You must discover some inherent inadequacy of the tools that you have been given. In fact, Abel and Galois did not prove that quintic polynomials have no solutions. Instead, they proved something more subtle: that the five operations listed above are inadequate to express the solutions, assuming they exist. Their proofs involved a very deep and novel exploration of the idea of symmetry.

Let's start with the original quintic polynomial:

$$x^5 + ax^4 + bx^3 + cx^2 + dx + f$$

Assuming that it has five roots, r_1, r_2, r_3, r_4, and r_5, then each of the coefficients of the original polynomial is a symmetric function of the roots. For example:

$$a = -(r_1 + r_2 + r_3 + r_4 + r_5),$$

$$b = r_1r_2 + r_1r_3 + r_1r_4 + r_1r_5 + r_2r_3 + r_2r_4 + r_2r_5 + r_3r_4 + r_3r_5 + r_4r_5,$$

and so on. Looking at these formulas, you may notice that each of the roots participates equally. More precisely, Galois observed that if you permute the roots in any way (e.g., by replacing r_1 with r_2 and r_2 with r_1), the expressions do not change. (The terms will be listed in a different order, but the sums will still be the same.). There are 120 different ways to permute five numbers and thus a typical quintic polynomial has 120 symmetries. However, it is worth noting that some polynomials have fewer symmetries. (The reason is technical, but some permutations may be forbidden because of extra algebraic relations between some of the roots—for example, one root might be the square of another.)

Abel realized, and Galois clarified, that if a polynomial is solvable by radicals, it creates a hierarchy of intermediate polynomials and a hierarchy of "number fields" corresponding to the roots of those polynomials. This is the reason for the nesting of radicals within radicals in Cardano's and Ferrari's formulas; each time you peel off a radical (like peeling the layers of an onion) you move to a lower number field. The symmetries of the original polynomial have to respect this hierarchical structure.

Now comes the difficult, but clinching point of Galois' argument. The full *group* (a term coined by Galois) of 120 permutations of the roots does not allow a tower of subgroups of the requisite type. It's as if you were trying to build a wedding cake 120 feet high; you can't do it. As it turns out, the maximum height (the maximum number of allowable permutations for a quintic polynomial to be solvable by radicals) is 20.

GALOIS' SOLUTION ACTUALLY provided a clear-cut criterion to determine which polynomials can and which ones cannot be solved by radicals. If you have a polynomial whose "wedding cake" (or Galois group) has 20 elements or less, you can solve it. Galois thought that his criterion was hopelessly impractical—but nowadays, thanks to the computer, the calculation of the Galois group can be automated. Thus, for example, the Galois group for the polynomial $x^5 - x + 2$ contains all 120 permutations, and therefore the solutions to the equation $x^5 - x + 2 = 0$ cannot be written in terms of the five algebraic operations.

The equation $x^5 - x + 2 = 0$ *does* have solutions. They just can't be expressed with the limited palette of $+$, $-$, $\times$, $\div$, and radicals. In 1858 Charles Hermite proved that the solutions to any quintic can be written down by using a new kind of function, called elliptic functions, which Abel had discovered.

This is a normal human response to a problem: If you can't overcome the difficulty with the tools you have, invent new tools. However, mathematicians

are not completely like ordinary people. Because their problems are often several steps removed from practical application, they often care more about how a problem is solved than whether it is solved. Abel's and Galois' proof that quintic equations cannot be solved by radicals has completely eclipsed Hermite's discovery of how they can be solved by elliptic functions.

But there is another reason for the enduring fame of Galois' proof. His concept of a group has now become the main tool that mathematicians use to express the ancient idea of symmetry. I find it very curious that the first explicit use of symmetry groups came in such a difficult context. It is as if no one had ever invented the wheel until the Wright brothers incidentally came up with it as a way to get airplanes off the ground. We would say, "Wow! Someone should have come up with that earlier!"

The idea of a symmetry group should be one of the most basic things in mathematics. In fact, the ability to perceive a symmetric object may even precede our ability to count. Perhaps its very obviousness made it difficult for mathematicians to discover. They could not formalize the meaning of symmetry until they encountered it in a context (the solution of polynomials) where its meaning was so far from obvious.

Fittingly, Galois' legacy was every bit as revolutionary as the political causes that he fought for. The tool he invented, group theory, has more than fulfilled the visions of its creator. Chemists now use group theory to describe the symmetries of a crystal. Physicists use it to describe the symmetries of subatomic particles. In 1961, when Murray Gell-Mann proposed his Nobel Prize-winning theory of quarks, the most important mathematical ingredient was an eight-dimensional group called SU(3), which determines how many subatomic particles have spin $1/2$ (like the neutron and proton). He whimsically called his theory "The Eightfold Way." But it is no joke to say that when theoretical physicists want to write down a new field theory, they start by writing down its group of symmetries.

15
the geometry of whales and ants
non-euclidean geometry

At the same time that a revolution was going on in algebra, similar events were taking place in geometry. Two millennia earlier, Euclid had written down a short set of axioms from which, supposedly, all of geometry could be derived. These axioms were intended to be self-evident truths that did not require any proof.

For centuries Euclid's *Geometry* was considered the *ne plus ultra* of deductive reasoning. The eighteenth-century philosopher Immanuel Kant built up a theory of knowledge, in which he cited Euclid's geometry as an example of "synthetic *a priori*" truth—in other words, infallible knowledge about the universe that is derived from pure reason rather than observation.

However, one axiom had always appeared a little bit clumsier than the others. The axiom in question is the "Parallel Postulate," which Euclid does not use until late in his first book: "If a straight line falling on two straight lines makes the interior angles on the same side less than two right angles, the two straight lines, if produced indefinitely, meet on that side on which are the angles less than two right angles." This assumption is used, for example, to prove that the sum of the angles of a triangle equals 180 degrees.

Many mathematicians felt the Parallel Postulate was true but far from self-evident, and thus a flaw in Euclid's otherwise sterling system of axioms. They

$$ds^2 = d\frac{x^2 + dy^2}{y^2}$$

dx and *dy* represent the sides of an "infinitesimal" triangle, and *ds* represents their hypotenuse.

took up the challenge of proving it from the other axioms that Euclid had provided. This mathematical grail quest lured the famous and obscure alike. Legendre (whom we have met already) believed that he had proved it. So, at one time or another, did less-famous mathematicians like John Wallis, John Playfair, Girolamo Saccheri, Johann Lambert, and Wolfgang Bolyai. In all cases, they made hidden assumptions that, under the harsh light of scrutiny by other mathematicians, were no better motivated than Euclid's postulate.

In the first half of the nineteenth century, three men separately and independently dared to think the unthinkable. Perhaps a valid geometry might exist in which the Parallel Postulate was actually false. This would be a non-Euclidean geometry—that is, a geometry in which one of the axioms laid down by Euclid, more than two millennia earlier, is expressly violated.

This idea was just as heretical as Hamilton's idea of an algebra with no commutative law. However, denying the Parallel Postulate took perhaps even more courage, because it had the great weight of Euclid, Kant, and two thousand years of tradition behind it.

The first of the three revolutionaries was Karl Friedrich Gauss, the most famous mathematician of his era. Gauss, a friend of Bolyai from their student years, dabbled at proving the Parallel Postulate in the early 1800s.

But gradually, around 1820, he seems to have become convinced that an alternative, non-Euclidean geometry could be constructed. However, he never published this idea, and only alluded to it somewhat vaguely in letters. The best evidence of his reasons comes from a letter he wrote in 1829 to his friend, Friedrich Bessel, in which he says that he feared the "howl from the Boeotians" (a pejorative term for stupid people) that would ensue if he published his work.

THE SECOND DISCOVERER of non-Euclidean geometry was Janos Bolyai, the son of Gauss's old school chum. Wolfgang, who became a mathematics teacher in Hungary, tried to warn his son against trying to prove the Parallel Postulate: "For God's sake, I beseech you, give it up. Fear it no less than sensual passions because it, too, may take all your time, and deprive you of your health, peace of mind, and happiness in life." But his son ignored the advice, and he eventually wrote a 24-page treatise on what he called the "absolute science of space," which his father generously published as an appendix to one of his textbooks in 1832.

The elder Bolyai naturally sent a copy to his old friend Gauss, who responded in unexpected fashion: "To praise [this work] would amount to praising myself. For the entire content of the work, the approach which your son has taken, and the results to which he is led, coincide almost exactly with my own meditations … It was my plan to put it all down on paper eventually, so that at least it would not perish with me. So I am greatly surprised to be spared this effort, and am overjoyed that it happens to be the son of my old friend who outstrips me in such a remarkable way."

In spite of the compliment at the end, it was a crushing blow to the younger Bolyai. Gauss was saying that his discovery of non-Euclidean geometry was nothing new. Janos never published another mathematical paper in his life. Not only had Gauss lacked the courage to publish the discovery himself, he had now compounded his mistake by discouraging an aspiring young mathematician who might have made a great name for himself.

Because Gauss was too reticent, and Bolyai gave up too easily, the third discoverer of non-Euclidean geometry deserves the most credit for bringing it to the world's attention. He was Nikolai Ivanovich Lobachevsky, a Russian mathematician who lived in Kazan, the ancient capital of the Tatars. He first published his version of non-Euclidean geometry in 1829 in a very obscure Russian journal, but unlike Bolyai he continued to write articles and books about it and finally succeeded in getting an article into *Crelle's Journal* in 1837. Even so, he did not receive the kind of acclaim during his lifetime that one might expect. Today, however, Lobachevski is considered one of the first great Russian mathematicians, and in Russia his geometry is called *Lobachevskian*. Western mathematicians call it, more descriptively, *hyperbolic geometry*.

What exactly is hyperbolic, or Lobachevskian, geometry? I think that the best way to think about it is to forget all about the Parallel Postulate and about Euclid. You must especially forget about the prejudice that you have surely been brought up with, that Euclidean is the "natural" geometry of the real world. Hyperbolic geometry is no more artificial than Euclidean. Think of it as the geometry of the ocean. If whales had invented geometry, the geometry they would have invented would be hyperbolic.

Suppose, for a moment, that you are a whale. Light is not very useful in the deep ocean, because the water is dark. So you mostly communicate and experience the world through sound. The shortest distance between two points in your world would be the path taken by sound waves. To you, this would be the analogue of a straight line.

Now here's the catch. Sound does not travel at a constant speed in the ocean. Below a certain depth, roughly 2000 feet (600 meters), it travels at a speed that is proportional to the depth below the surface. So the path that sound waves travel is not straight, but curved. A sound wave will get from whale A to whale B quicker if it goes downward, to exploit the greater sound

---

Below Demonstration of the curves along which sound travels in the ocean.

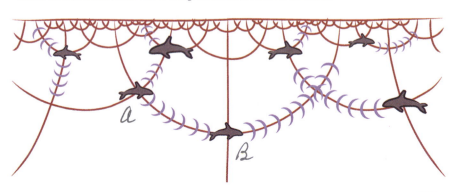

speed at depth, and then comes back up. In fact, we can be more precise about the nature of these curves: they are arcs of circles centered at the ocean surface! Thus, to a whale, what humans call a "circle" is actually a "line" (the shortest distance between two points).

Whale Geometry is a geometry where some surprising (to us) things happen, but they would not be the least bit surprising to whales. The sum of the angles of a triangle is less than 180 degrees. Rectangles (four-sided figures with all right angles) do not exist; however, right-angled pentagons do. Most importantly, it is a geometry of *negative curvature*. This means that lines that start out parallel tend to move farther and farther apart.

AMAZINGLY, ANOTHER non-Euclidean geometry, besides hyperbolic geometry, had been known for centuries—only no one ever thought of it in those terms. It is the geometry of a sphere. On the surface of a sphere (such as Earth), the sum of the angles of a triangle is greater than 180 degrees. Rectangles do not exist, but right-angled triangles do. Keep in mind the curvature of the Earth! For example, a triangle can be drawn with three right angles: start at the North Pole, travel in a straight line down to the Equator, then travel due east or west a quarter of the way around the globe, and then go due north again. You will trace out a triangle with three 90-degree angles. Spherical geometry is a geometry of positive curvature. In other words, lines that start out parallel (such as meridians, near the Equator) tend to move closer and closer together, and they eventually converge at the poles.

The reason that no one ever thought of spherical geometry as an alternative to Euclidean geometry is simple: We can see a sphere as being imbedded in three-dimensional Euclidean space, so its "non-Euclidean-ness" is not immediately obvious. Suppose, however, that you were unable to perceive a third dimension beyond the surface of the sphere. For example, perhaps you are an ant, living on the surface of an asteroid with no oceans (so you can go anywhere you want to). You have no concept of space, no concept of underground;

Left Spherical geometry and the curvature of the Earth.

everything you know is the surface of your spherical world. The curvature of that world is positive and its geometry is non-Euclidean. We could call it Ant Geometry.

Instead of one geometry of nature, we can now see there is a whole spectrum of geometries with different amounts of curvature, ranging from Ant Geometry (spherical) to Human Geometry (Euclidean) to Whale Geometry (hyperbolic). But that's not all. These are only the geometries of *constant* curvature. We can also imagine geometries whose curvature varies from place to place. They can be two-dimensional, three-dimensional, or even higher. Gauss (perhaps influenced by his unpublished thoughts on hyperbolic geometry) was the first mathematician to understand the concept of varying curvature in a two-dimensional space, and his student Bernhard Riemann extended the concept to higher dimensions in 1854. Both of them thus anticipated one of the epochal discoveries of the twentieth century: Albert Einstein's theory of general relativity, which postulates that our four-dimensional spacetime has curvature that varies from place to place. Without Lobachevski, Bolyai, Gauss, and Riemann, Einstein would never have been able to write down the equations for his theory.

Above An engraving displaying an "Allegory of Geometry," by F. Floris, 16th century.

16

in primes we trust
the prime number theorem

Gauss's mishandling of the discovery of non-Euclidean geometry was one of the few black marks on an otherwise remarkable career. He contributed so much to so many parts of the subject, and he virtually created the modern subject of number theory, which deals with the properties of whole numbers and especially with the solution of equations in whole numbers.

Gauss was born in 1777 in Brunswick, Germany. As a child prodigy, Gauss attracted the attention of the Prince of Brunswick, who supported him through preparatory school and the University of Göttingen, where he earned his doctorate in 1799 with his first, not entirely satisfactory proof of the Fundamental Theorem of Algebra. (He later gave three more proofs.)

Gauss always had a special place in his heart for number theory, a subject he called the "queen of mathematics." His earliest significant discoveries came in this subject. In 1796, still a student at the university, Gauss proved that a regular 17-sided polygon can be constructed by a ruler and compass—a discovery that had eluded the ancient Greeks, who first took an interest in such construction problems. Although it looks like a theorem of geometry, this theorem is closely linked to the solvability of polynomials.

The key question is whether the angle $(360/17)°$ can be constructed with ruler and compass. If so, the 17-gon can be constructed simply by piecing

$$\pi(n) \approx \int_{2}^{n} \frac{dx}{\ln(x)}$$

The function $\pi(n)$ [not to be confused with the number π] represents the number of primes less than n. The prime number theorem says that this total is *roughly* equal to the integral of a density function, $1/\ln(x)$. Though it is only approximate, the formula gets more and more accurate (on a percentage basis) as n gets larger.

together 17 isosceles triangles with this angle at their vertex. As long ago as 1637, in his book *La Géometrie* (which Gauss had surely studied), René Descartes had found a simple criterion for constructability of a line segment. Namely, a segment can be constructed with a ruler and compass from a given line segment of unit length if its length can be expressed using only whole numbers and the five algebraic operations $+$, $-$, $\times$, $\div$, and $\sqrt{\ }$. This should look somewhat familiar—it looks a lot like the toolbox for solving polynomial equations by radicals. But it is more restricted, because only square roots are allowed—no third or higher roots. Likewise, an angle is constructible if its cosine and sine are constructible lengths.

Gauss's audacity is amazing. To prove that $(^{360}/_{17})^{\circ}$ is a constructible angle, he solved the polynomial equation $x^{17} = 1$. At this point, in 1796, no one knew whether degree-five equations were solvable, even using cube roots, fourth roots, and fifth roots. Gauss was proposing to solve a degree-*seventeen* equation, with fewer tools. And he succeeded!

Five years later, in 1801, Gauss published his first book, *Disquisitiones arithmeticae*. It was the first systematic book on number theory, establishing its methods and identifying its interesting questions. His theorem on 17-gons appears there, along with a generalization: an *n*-sided polygon

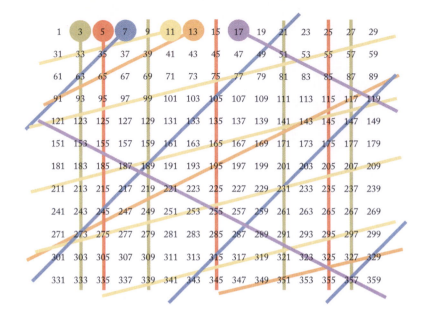

1	3	5	7	9	11	13	15	17	19	21	23	25	27	29
31	33	35	37	39	41	43	45	47	49	51	53	55	57	59
61	63	65	67	69	71	73	75	77	79	81	83	85	87	89
91	93	95	97	99	101	103	105	107	109	111	113	115	117	119
121	123	125	127	129	131	133	135	137	139	141	143	145	147	149
151	153	155	157	159	161	163	165	167	169	171	173	175	177	179
181	183	185	187	189	191	193	195	197	199	201	203	205	207	209
211	213	215	217	219	221	223	225	227	229	231	233	235	237	239
241	243	245	247	249	251	253	255	257	259	261	263	265	267	269
271	273	275	277	279	281	283	285	287	289	291	293	295	297	299
301	303	305	307	309	311	313	315	317	319	321	323	325	327	329
331	333	335	337	339	341	343	345	347	349	351	353	355	357	359

is constructible if all of the odd prime factors of n are one greater than a power of 2, and furthermore are raised to only the first power. Only five such primes are known: 3 (2^1 + 1), 5 (2^2 + 1), 17 (2^4 + 1), 257 (2^8 + 1), and 65,537 (2^{16} + 1). This result is as far from being practically useful as any theorem can be. It would probably take a lifetime to perform the construction of a 65,537-sided polygon, and when you finished it would be impossible to tell the result apart from a circle!

Above A diagram showing a method for identifying prime numbers, described by the ancient Greek mathematician Eratosthenes (276 BC– 194 BC).

This example gives an inkling of the central role of prime numbers in number theory. These are the numbers that combine (by multiplication) to form all others, and in this sense they are as fundamental as the elements in chemistry. They are important both as a tool for solving other problems, and as a subject of study in their own right. One of their enduring mysteries is to understand how they are distributed.

Here is the paradox. Primes behave very much as if they were randomly distributed on the number line. The distribution is not completely uniform; large numbers are less likely to be prime than small numbers, because there are more possible prime divisors. Gauss conjectured on the basis of empirical evidence that the "density" of prime numbers decreases in proportion to the natural logarithm of n, written $\ln(n)$. This means that a ten-digit number is half as likely to be prime as a five-digit number, and five times less likely to be prime than a two-digit number. We can think of $^1/_{\ln(n)}$ as the "probability" that n is prime. And yet this statement is absolutely paradoxical, because there is no probability involved! Either a number is prime or it is not.

Nevertheless, Gauss's conjecture, called the Prime Number Theorem after it was finally proved in 1898, provides remarkably accurate estimates of the distribution of primes. For example, the density formula says that the number of primes less than 1 million should be about 78,628. In reality, the number of primes is 78,498—an error of less than 0.2 percent. If we go up to 1 billion, the predicted number is 50,849,235. The exact number is 50,847,534—so the estimate is off by less than 0.004 percent! I hope that you are as amazed by this fact as I am. Think of how difficult it is to determine whether even *one* large number is prime. Even with current computer technology, no one can tell whether a randomly chosen 200-digit number is prime. And yet, using the Prime Number Theorem, we can find a very accurate (though not perfectly accurate) count of *all* the primes less than that number!

THE STORY OF the Prime Number Theorem is a little reminiscent of Fermat's Last Theorem. At some unknown date, Gauss wrote the cryptic comment, "Prime numbers less than $a \approx {}^a/_{\ln a}$," which is roughly a statement of the Prime Number Theorem. There is no indication of a proof, and he probably based his assertion on numerical evidence. Around 1850, the Russian mathematician Pafnuty Chebyshev proved that the error in the above approximation, for large enough numbers n, is never greater than 11 percent. Of course, the examples above intimate that the error is in fact considerably smaller. Chebyshev's work was a big step in the right direction, in part because he used the zeta function as a tool for counting the number of primes.

In 1859, Bernhard Riemann took another amazing step forward, which explains why the zeta function (discussed in Chapter 12) is now named after him rather than Chebyshev. He discovered an exact formula for the number of primes less than n. However, there is a catch. To compute the number exactly, you need to know the infinitely many places in the plane where the Riemann zeta function takes the value zero. (These places are called the "zeros" of the zeta function, as shown below). If you know approximately where the zeros are, Riemann's formula tells you approximately how many primes there are.

In 1898, Jacques Hadamard and Charles de la Vallée Poussin, working separately, both proved that all the zeros lie in an infinite strip, to the right of the line $x = 0$ and to the left of the line $x = 1$. Even this rough information on the location of the zeros was good enough to prove the Prime Number Theorem, one of the landmark theorems of the nineteenth century. Fortunately for Hadamard and de la Vallée Poussin, Gauss was no longer alive to say, "I knew that 100 years ago!"

The story of the theorem is not quite over, though. The more accurately you can pin down the location of Riemann's zeros, the more you know about prime numbers. Hadamard and de la Vallée Poussin showed the zeros lie on an infinitely long and straight "street" in the plane (the shaded strip in the figure below). Riemann had conjectured, but could not prove, a much more precise statement: All of the zeros lie exactly in the middle of this street! If this statement, called the "Riemann Hypothesis," is true, it would provide the finest control over the distribution of primes.

Now that Fermat's Last Theorem has been proved, the Riemann Hypothesis is at the top of number theorists' "most wanted list." In 2000, the Clay Mathematics Foundation named it one of seven "millennium problems," and offered

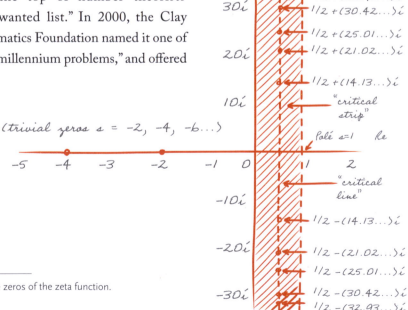

Right The zeros of the zeta function.

a reward of a million dollars for its solution. Because it is so technical, there are few elementary examples of problems that the Riemann Hypothesis would solve. However, here is one example.

Back when I was in second grade, I noticed that the decimal expansions of certain fractions take a long time to repeat, while others do not. For example, $1/3 = 0.3333\ldots$ is a quick repeater, as is $1/37 = 0.027027\ldots$ On the other hand, $1/7 = 0.1428571428\ldots$ is a slow repeater, cycling through six digits before it starts over. An even slower repeater is $1/19 = 0.0526\ldots$, which goes a full 18 digits before starting over with $\ldots0526\ldots$

C. F. Gauß

In fact, the decimal expansion of any number $1/n$ will eventually settle into a repeating cycle of no more than $(n-1)$ digits. The numbers that take the full $(n-1)$ digits are always prime, like 7 and 19. However, not all primes are slow repeaters. For example, $1/37$ starts repeating long before the 36th digit—it takes only three digits! There is no known formula to determine which prime numbers are fast repeaters or slow repeaters. However, if the Riemann Hypothesis is true, then about 37.4 percent of all primes are slow repeaters. This is typical of the amazingly precise information about primes that number theorists can squeeze out of the Riemann Hypothesis. (This result was proved by Christopher Hooley in 1967.)

Above Carl Frederich Gauss (1777–1855).

How good is the evidence for the Riemann Hypothesis? To date, ten trillion zeros of the zeta function have been found, and they all lie exactly in the middle of the "critical strip," just as Riemann predicted. Any reasonable scientist, in any other subject, would have declared the problem solved long ago. However, in such matters mathematicians are not reasonable.

17
the idea of spectra
fourier series

So far, the stories I have told in this chapter have not portrayed French mathematics in a very positive light. First the French Academy of Sciences lost Abel's memoir on elliptic functions, and then it spurned Galois' revolutionary discovery of group theory. Some of the most unexpected breakthroughs in the first half of the 1800s came from mathematicians elsewhere—Hamilton in Ireland, Abel in Norway, Bolyai in Hungary, Lobachevsky in Russia, and of course Gauss in Germany. Nevertheless, the center of mathematics at this time was undoubtedly in Paris. Any student of mathematics today will inevitably encounter a whole suite of French names from this period, including Lagrange, Laplace, Legendre, Cauchy, Liouville, Poisson, Fourier.

It is remarkable that French mathematics remained so strong throughout a period of huge political upheaval. It weathered the French Revolution, the Terror, then Napoleon's rise, his exile and return, the restoration of the monarchy, the abdication of King Charles and ascension of King Louis-Philippe, and finally the Second Republic and Second Empire. All of these swings of political fortune had repercussions for individual mathematicians, whose fortunes rose and fell with their leaders of choice. Nevertheless, French mathematical culture as a whole prospered. Perhaps one reason

$$\hat{f}(n) = \frac{1}{2\pi} \int_{-\pi}^{\pi} f(x) e^{inx} dx$$

$$f(x) = \sum_{n=-\infty}^{\infty} \hat{f}(n) e^{inx}$$

For any function $f(x)$, "f-hat" is its Fourier series. It decomposes f into a spectrum of sine and cosine waves of different frequencies. The second formula tells how to reconstruct the original function f from its spectrum. In a sense, it says that "f-hat-hat" equals f again.

was France's increased social mobility, which gave access to education and professional opportunities to anyone with talent and a little bit of luck.

A PERFECT EXAMPLE would be Joseph Fourier, a tailor's son who was born in 1768 and orphaned at age nine. Brought up in a convent and educated in a military school, he supported the Revolution and managed not to be executed during the Terror, although he was arrested twice. He rose through France's top educational institutions, the École Normale where he was a student, and the École Polytechnique where he became a junior professor.

In 1798, Napoleon Bonaparte launched a military campaign in Egypt, with the idea of making it into a French colony. In addition to his invasion force of 40,000 soldiers, Napoleon brought 167 scientists to study Egypt and catalog Egyptian culture.[†] Among the "savants" who came along was Fourier.

---

†. In spite of his other failings, Napoleon was an admirer and supporter of science. He even has a minor theorem in Euclidean geometry, Napoleon's theorem, named after him. It is unclear whether Napoleon actually proved it.

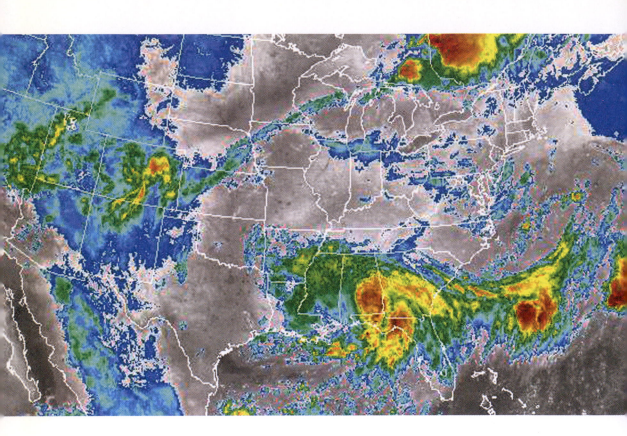

Above The heat equation has numerous practical uses, including weather forecasting.

This was apparently the first time Fourier met Napoleon, and the association would change his life. The military campaign was a disaster—the British navy destroyed the French fleet after Napoleon reached Egypt, stranding his massive army—but Fourier evidently made a good impression on the future emperor. After Fourier returned to France in 1801, Napoleon appointed him prefect of Isère, a province on the Italian border. Fourier was not entirely happy about this, as he would have preferred to stay in Paris at the École Polytechnique, but he proved to be a capable public servant. Napoleon's defeat at Waterloo in 1815 ended Fourier's political career, but it actually helped his scientific career. He moved back to Paris, where he became the secretary of the Academy of Sciences in 1822 and died in 1830.

Like Abel and Galois after him, Fourier struggled to obtain recognition for his most important work, but for a different set of reasons. Beginning in 1802, he began to conduct experiments on the diffusion of heat in solid materials. He began with very simple cases—first a solid bar, then a ring—which could be treated as one-dimensional problems. At the same time, he developed a two-part mathematical theory of these objects, first setting up an equation (known as the heat equation) that expresses the conduction of heat inside the bar, and then solving it by a method that became known as Fourier series.

THE HEAT EQUATION is an excellent example of what mathematicians in the ninetheenth century did: it indicates exactly how the current temperature distribution affects the future temperature. Roughly, it says that heat will flow toward points that are cooler than the average temperature of their neighbors, and away from points that are warmer. Because this is a statement about rates of change, of course it is expressed in the language of calculus. Furthermore, it relates two different kinds of rates of change. The rate of change in temperature over time, written in the formula as du/dt is determined by the spatial variations in temperature, represented by d^2u/dx^2, which reflects the difference between the temperature at the point x and the temperature at two equally spaced points to its left and right. The complete heat equation reads as follows:

$$\frac{\partial u}{\partial t} = k \frac{\partial^2 u}{\partial x^2}$$

Such an equation is called a partial differential equation: "partial" because each term expresses part of the way that temperature varies (either in space or in time); "differential" because it involves derivatives. Partial differential equations would turn out to be crucial for modeling all sorts of physical processes, from heat conduction to fluid flow to the propagation of electric and magnetic fields. Every time you read a weather forecast, you are seeing the solution of several partial differential equations that describe the motion of heat and air and water in the atmosphere.

Fourier's work also illustrates the fact that mathematics, when applied to real-world problems, is a two-step procedure. First comes the modeling of the problem—translating your assumptions, or your empirical observations, into mathematical language. Fourier's modeling of heat flow is beautiful, convincing, and far-reaching. The three-dimensional heat equation applies to everything from the inside of your coffee cup to the inside of a star to global climate change.

The next step after modeling is to solve the equations of the model. It would seem that this would be the most routine part of the work—a solution is a solution, it is either correct or not—and yet this was exactly where Fourier ran into controversy.

Fourier used a time-honored method of solving the equation: he guessed. In particular, because the temperature u in the bar is a function of both space (x) and time (t), he guessed that it was simply a product of two functions, one of them purely a function of time and the other one purely a function of space. It worked; the solution was a product of a sine wave (in space) and a decaying exponential function (in time). If your metal bar starts with a temperature distribution whose graph is a sine wave, its temperature will gradually cool down to zero (or whatever the ambient temperature is) at a rate that is proportional to the square of the wavelength of the sine wave.

But what if the initial temperature distribution of your metal bar *isn't* a sine wave? For example, in his experiments Fourier put one end of the bar into a furnace, creating a temperature distribution with half of the bar hot

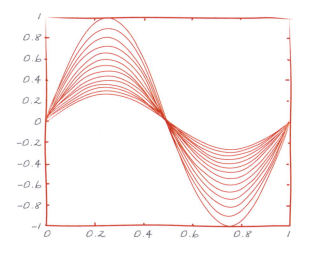

Above Fourier's solution for the heat equation involves sine waves whose amplitudes decrease over time, as shown here.

and half cold. In physics lingo, this would be called a "square wave," not a sine wave. But Fourier asserted that *any* temperature distribution could be written as a sum (not just a finite sum—an *infinite* sum, nowadays called a Fourier series) of sine waves.

Nowadays, with computers, we can draw beautiful pictures to illustrate Fourier's idea of approximating arbitrary functions with trigonometric series. In particular, it is easy to see how a square wave emerges out of a chorus of wobbly approximations. But this precise point stuck in the throats of his colleagues, particularly his former teacher Joseph Louis Lagrange. It implied a sea change in mathematicians' conception of what a function was.

EVER SINCE EULER, functions had been seen as formulas: finite combinations of known functions such as polynomials, exponentials, trigonometric functions, and so forth. Or, following Newton, they had been expressed as power series, which are basically "infinite polynomials." But Fourier series were much more versatile. They could represent functions with jumps and corners, which could not be expressed with simple arithmetic formulas. Fourier's paper marked the beginning of a broader conception of a function, the input-output model that we use today. A function is simply

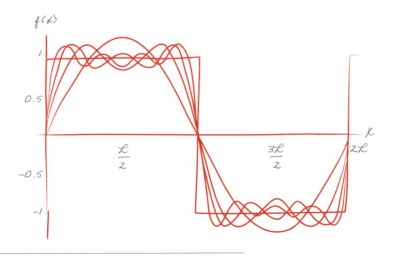

Above A graph displaying the "square wave" and its approximation by finite sums of sine waves. The approximations can be made as close as desired to the original square wave.

a rule that assigns to any input value a unique output. The input and output values don't even have to be real numbers, and the rule certainly does not have to be expressible as a formula. In Part Two, I said perhaps a bit cavalierly that classical mathematicians were not interested in quantities like heartbeats and stock prices. It would be more accurate to say that it wouldn't even have occurred to them to think of such things as mathematical functions. Fourier's insight opens the door to a vast range of physical and empirical processes, especially ones with jumps and discontinuities.

Lagrange's objections did have some merit, though. Fourier said that you could break any function down into a sum of sine waves, each with a different frequency n. There is a function $\hat{f}(n)$ or "f-hat" that tells you how "strong" each frequency is. Fourier's key point is that you can reconstruct the original f from "f-hat." According to Fourier's inversion formula, "f-hat-hat" is equal to f again. Fourier did not provide an adequate proof of this. In fact, it is not even true for functions with discontinuities. What kinds of functions do obey the Fourier inversion formula? The answer is not easy, and the problem provided a major stimulus in the nineteenth and twentieth centuries to the theory of functions, or "functional analysis."

Above Joseph Fourier (1768–1830).

THE IMPORTANCE OF the Fourier series (and the "hat" concept, which is technically called a Fourier transform) goes far beyond the heat equation. Fourier transforms allow any time-varying signal to be decomposed into a *spectrum* of wavelengths. Astronomers use this principle to determine what molecules are in distant stars. Radios use this principle to pick out a particular channel—it's a matter of finding a particular wavelength in a time-varying signal. Music synthesizers use Fourier series to simulate the sound of a violin or a flute, or to create a new sound that has never been heard before. In other words, they are tweaking "*f*-hat" in order to produce, hopefully, a better-sounding "*f*." Fourier series and transforms are all around us; we just don't know it.

As for Fourier, he had to wait a long time to see his paper published. He presented it to the Institute of France in 1807, and it was rejected because of Lagrange's objections. (Also, another academician named Jean-Francois Biot complained that Fourier should have given him more credit.) Fourier submitted a re-worked version for a prize in 1811 and it won, but Lagrange still deemed it unsuitable for publication. Finally, in 1822, with Lagrange dead and Fourier now installed as secretary of the Academy, his treatise *The Analytic Theory of Heat* finally appeared, and it became one of the most widely read mathematics books of the nineteenth century.

18

a god's-eye view of light
maxwell's equations

While mathematics was experiencing revolutions in algebra, geometry, and the theory of functions, physics was undergoing its own revolution.

At the beginning of the nineteenth century, the theories of mechanics and gravity were in pretty good shape. Newton had explained how planets orbit around the Sun. Euler, Laplace, and others had explained multiple-body interactions in the solar system, such as the precession of the equinoxes and the slow variations in Jupiter and Saturn's orbits. Newton's laws had explained how solid objects respond to mechanical forces, and Euler's equations of hydrodynamics had done the same thing for fluids.

However, three subjects in physics remained entirely mysterious to the scientific community: electricity, magnetism, and the nature of light. As of 1800, there was not the slightest bit of evidence that any of these three phenomena were related to the others. Yet by 1865, that had all changed and physicists had arrived at a theory that unifies all three subjects. Magnetic fields are produced by electric currents. Electric fields are generated by changing magnetic fields. And light is nothing more than a traveling electromagnetic wave—an intricately woven tapestry of vibrating magnetic fields and electric fields that cross one another like the warp and the weft of a piece of fabric.

$$\nabla \cdot \vec{E} = 0$$

$$\nabla \cdot \vec{B} = 0$$

$$\nabla \times \vec{E} = -\frac{\partial \vec{B}}{\partial t}$$

$$\nabla \times \vec{B} = \frac{1}{c^2} \frac{\partial \vec{E}}{\partial t}$$

E and *B* represent the electric and magnetic fields in a vacuum, with no electric charges or currents present. The constant *c* is the speed of light. The symbol "∇" (the divergence) represents the tendency for field lines to move apart. The symbol "$\nabla \times$" (the curl) represents the tendency of the field lines to rotate. Collectively, the equations say that in the absence of electric charges, neither the electric field nor the magnetic field has any sources or sinks.

In order to reach these conclusions, physicists first had to assimilate a number of startling experimental discoveries. Then they had to develop a new kind of physics, in which solid, tangible objects (like wheels, bars, pulleys, and levers—the stuff of mechanics) were replaced by intangible concepts such as electric and magnetic fields. Because common sense and everyday experience no longer apply to these intangible but real phenomena, physicists were forced to embrace mathematics in a deeper way than they ever had before. It was the only guide that worked when intuition and our senses failed.

THE NATURE OF LIGHT had been debated as early as the 1600s, when Isaac Newton argued that it consisted of tiny corpuscles, while Robert Hooke insisted that it was made of waves. Newton's enormous prestige pushed the wave theory into the background for a hundred years or so. But in the early 1800s, several experimental discoveries revived the debate. In 1801, Thomas Young discovered the interference of light waves. When a beam of light passes through two narrow, parallel slits, what we see on the other side is not two narrow bright bands, but a series of alternating dark

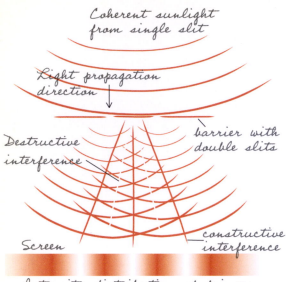

Coherent sunlight from single slit

Light propagation direction

Destructive interference

barrier with double slits

Screen

constructive interference

Intensity distribution of fringes

and light bands with the brightest one right in the middle. This is easy to explain if you think of light as being like ripples of water in a tank, but not if you think about it as tiny particles of grapeshot.

Also, as early as 1665, Francesco Grimaldi had observed an effect he called diffraction—the apparent bending of light around a corner. Again, this was hard to square with Newton's laws. (Remember that particles in motion with no force acting on them are supposed to go in a straight line.) Refraction, the bending of light as it passes through a prism, was also easier to explain with the wave theory than the particle theory. In 1818, Augustin Fresnel successfully accounted for all three of these phenomena—interference, diffraction, and refraction—with a theory in which light consists of transverse waves.[‡]

By the 1820s in France, and the 1830s in England (which was slower to shake off its hero-worship of Newton), the wave theory had gained the upper hand. But if light was a wave, what was the wave made of? It could not be a wave of air or any other fluid, because transverse waves don't travel through fluids; they require a medium with elasticity, or the ability to "snap back" after being stretched. The great majority of physicists assumed that light traveled through some sort of "luminiferous aether," but all efforts to detect this aether directly were in vain.

Meanwhile, the mysteries of electricity and magnetism were also deepening. In 1799, Count Alessandro Volta of Italy had invented the battery, which for the first time made it possible for physicists to experiment

---

[‡] A transverse wave is one that propagates at right angles to the motion of the individual particles in the wave. An example is "the wave" (sometimes called "the Mexican wave") at a sports stadium. The individual particles (i.e., the fans) go up and down, but the wave moves around the stadium.

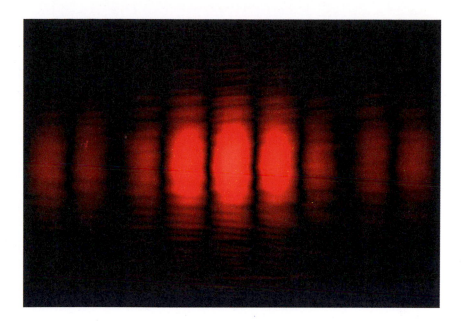

with steady electric currents. In 1820, Hans Christian Ørsted noticed, while preparing for a lecture, that when he turned on an electric current in a wire, it deflected a nearby compass needle. This was the first indication that electricity and magnetism were related. This clue was followed in 1831 by Michael Faraday's discovery of electromagnetic induction. Faraday showed that a changing electric current in one coil would induce a temporary electric current in another one. Likewise, moving a magnet close to a coil would temporarily induce a current. This was, then, a reciprocal effect to the one Ørsted had noticed. Magnetism could induce electricity, but only if the strength of the magnetism was changing.

THE MAN WHO WOVE all of these confusing clues into a beautiful theory was James Clerk Maxwell, a Scottish physicist. For those who think that great discoveries are always made in a flash of inspiration—like William Rowan Hamilton's discovery of quaternions—Maxwell provides compelling evidence to the contrary. He worked on electromagnetism for several years, gradually painting the beautiful canvas we now know as Maxwell's equations.

Maxwell's first step, in 1855, was to take seriously Faraday's description of the "lines of force" created by a magnet—lines that are easily seen if you sprinkle iron filings nearby. Faraday believed that the space around the magnet was surrounded by these "lines of force" even when no iron filings were present. Maxwell gave this invisible collection of curves a name—the magnetic field. He also postulated an electric field that conveys electric forces.

In the twenty-first century, we are completely accustomed to the idea that we live surrounded by electric and magnetic fields. So it may take a conscious effort to imagine how radical the idea was in the 1850s. What *is* an electric field? You can't see it. You can't touch it. How can you tell that it's there?

An additional roadblock to Maxwell's theory of fields was, again, the legacy of Newton. In Newton's theory of gravity, planets attract each other from a distance, with a force proportional to the inverse square of their distance. For a while, electricity and magnetism seemed to work in exactly the same way. Physicists subscribed to the idea of "action-at-a-distance" as an article of faith. But Faraday and Maxwell questioned this conviction. They said the force between two charges, or two magnets, results from the field between them. In Newton's universe, empty space is empty. But in Maxwell's universe, it is humming with electric and magnetic potential.

Opposite Iron shavings are used to reveal magnetic field lines produced by two bar magnets.

Six years after his first paper, Maxwell added another stroke to his scientific painting. He envisioned electricity as an elastic force in the medium that electric and magnetic fields inhabit. It's interesting to note that he had not yet abandoned the mechanical way of thinking and accepted the greater flexibility of mathematics. His second paper relies on an extremely complicated model, replete with spinning vortices to represent the magnetic fields and counterrotating "idle wheels" to represent the electric fields. All of this baroque machinery would be discarded in his third paper.

Elastic forces, as noted above, are exactly what is necessary to transmit transverse waves. Not only that, there is a simple formula for the speed of waves in any elastic medium. Reasoning by analogy, Maxwell was led to a formula for the speed of an electromagnetic wave. At the time, he was spending the summer at his estate in Scotland, and he could not look up the necessary physical constants to plug into the equation. But when he got back to his office at Kings College in London in the fall of 1861, he computed the speed as 310,740,000 meters per second. By comparison, in 1849, a French physicist named Armand Fizeau had measured the speed of light at 314,850,000 meters per second! (The currently accepted value is 299,792,458 meters per second. In fact, since 1983 the meter has been defined as the distance light travels in $1/299{,}792{,}458$ of a second, so the speed of light is now prescribed by definition and is no longer an experimental constant.) It could not be an accident, thought Maxwell, that the two constants were so close. In his paper announcing the result, he wrote in italics: "*We can scarcely avoid the inference that light consists in the transverse undulations of the same medium which is the cause of electrical and magnetic phenomena.*"

BUT MAXWELL was not done. Having used a mechanical analogy to discover that electromagnetic waves and light waves are the same thing, he realized that he could forget about the vortices and the counterrotating gears, and derive the result solely from mathematics. What was left, by the time he wrote his third paper in 1865, was a simple set of four partial differential equations that relates the electric field ($\mathbf{E}$) to the magnetic field ($\mathbf{B}$) at any point in a vacuum.

By themselves, these equations are not a complete theory of electromagnetism. In particular, they lack any information on how material particles, such as electric charges and magnets, respond to the fields $\mathbf{E}$ and $\mathbf{B}$. To use a Judeo-Christian analogy, Maxwell's equations represent the state of the universe after God said "Let there be light," and before he created anything else. To incorporate the material world, Maxwell added extra terms (representing charge density and current density) and extra equations.

Most of Maxwell's equations were not actually original to him. The individual equations are known as Gauss's law, Faraday's law, and Ampère's law. Maxwell's only new contribution was a correction term that enters Ampère's law when electric currents are taken into account. Nevertheless, the understanding that the equations could be brought together into a system, and the idea that magnetic and electric *fields* were the fundamental agent, were entirely due to Maxwell. So, too, was the discovery that the speed of light, c—the only experimental constant to be found in these equations—is a fundamental physical law.

Earlier it was mentioned that Euler's equation $e^{i\pi} + 1 = 0$ was voted the most beautiful equation of all time by readers of *Mathematical Intelligencer*. In 2004, *Physics World* conducted a similar poll. It is no surprise that the readers of that publication chose Maxwell's equations as the greatest ever. They are so simple, so symmetric, so hard-earned, and they explain so much.

Yet like the other revolutions described in this chapter, they made little impression at the time. Maxwell's contemporaries just didn't know what to make of them. "As long as I cannot make a mechanical model all the way through I cannot understand; and that is why I cannot get the electromagnetic theory," said William Thomson, Lord Kelvin, in 1884 (the same Lord Kelvin who couldn't "get" quaternions!).

But over time, the significance of Maxwell's equations became more apparent. They predicted that electromagnetic waves could exist with different wavelengths—such as the waves we now call microwaves, infrared, ultraviolet, and X-rays. They predicted that such waves could be created by oscillating electric fields. In 1901, Guglielmo Marconi used precisely this principle to transmit the first radio waves. They implied that light itself can exert pressure. Sure enough, researchers in the twentieth century discovered the "solar wind," which explained the centuries-old mystery of why comet tails point away from the Sun. And in 1905, as will be discussed further in the next chapter, they led Albert Einstein to the theory of relativity.

equations in our own time

PART FOUR

No scientist is more

emblematic of the twentieth century than Albert Einstein. One part imp and one part prophet, he sticks out his tongue at us in one famous photograph, and looks at us with world-weary eyes in another. His unruly hair and lack of concern for social conventions have come to define the popular image of science. He was the world's first scientist-as-rock-star.

In some ways science could not have asked for a better ambassador to the public. Einstein's fame was truly deserved. Not just once, but over and over again, he transformed the worldview of physicists. He was the first physicist to understand the quantization of light; he was the first to recognize the equivalence of matter and energy; his name is synonymous with the theory of relativity. Yet he also transcended science. He used his fame to advance the cause of pacifism, at least until the rise of Nazi Germany made that position untenable for him. In 1940, he warned President Franklin Roosevelt about the threat posed by a possible atomic bomb, a warning that paved the way for the Manhattan Project and profoundly affected the balance of power in the postwar world.

Einstein's discoveries were just as much a result of his personality and the time in which he lived as of his intellect. By nature he loved to question authority. While other physicists were hesitant to discard centuries of tradition, Einstein was completely unconcerned and even happy to do so. He was lucky to come of age at a time when physicists had three mature, thoroughly understood theories—mechanics, electromagnetism, and thermodynamics—that fundamentally contradicted each other in subtle ways. While others averted their eyes from the contradictions, Einstein dared to look straight at them and pointed out how to overcome them.

Curiously, Einstein was not a lover of mathematics early in his career. His former math teacher Hermann Minkowski once wrote, "In his student days Einstein had been a lazy dog. He never bothered about mathematics at all." But Einstein's attitude completely changed over time. Minkowski's mathematical reformulation of special relativity, in 1908, helped his theory win acceptance. Einstein could never even have written down his theory of general relativity without a deep understanding of non-Euclidean geometry. By 1912, Einstein had recanted his former disdain: "I have gained enormous respect for mathematics, whose more subtle parts I considered until now, in my ignorance, to be pure luxury!"

Sometimes converts make the best missionaries. Even if Einstein was a reluctant mathematician, he certainly enhanced the prestige of the subject. Thus it makes sense to start our history of the equations of the twentieth century with him.

19
the photoelectric effect
quanta and relativity

The greatest revolution in twentieth-century physics began with a seemingly insignificant observation. In 1887, the German physicist Heinrich Hertz noticed that he could get a spark to jump between two metal electrodes more easily when the electrodes were exposed to light (specifically ultraviolet light) than he could if they were in the dark.

Another German physicist, Philipp Lenard, showed in 1902 that shining a light on a metal caused the metal to emit what were then known as "cathode rays," and are now known as electrons. If the electrons had sufficient energy, they could produce the sparks that Hertz had seen. The phenomenon became known as the photoelectric effect: the production of electricity from light. By varying the intensity and frequency (i.e., the color) of the light, Lenard discovered some strange things about the photoelectric effect. It does not occur at all with red light, no matter how intense the light is. Also, the energy of the emitted electrons does not increase when the *intensity* of the light increases—only when the *frequency* increases.

In 1905, Albert Einstein, who was at that time a 26-year-old patent clerk in Bern, Switzerland, proposed a revolutionary explanation for Lenard's discovery. He hypothesized that light "behaves like a discontinuous medium consisting of energy quanta." His "energy quanta" are now called photons.

$$\mathcal{E} = mc^2$$

E represents the energy of an object; *m* its mass at rest; and *c* = 299,792,458 meters per second is the speed of light. Einstein's equation implies that matter is a form of energy.

Einstein argued that each photon contains a characteristic amount of energy, which is proportional to the frequency, ν, of the light: $E = h\nu$.

The quantization of light means that the absorption of a photon is an all-or-nothing deal. When a photon hits the surface of the metal, it cannot be half absorbed and half reflected (as an ordinary wave can). If it is absorbed, *all* of its energy goes into the target. If that energy exceeds the binding energy (P) that holds electrons to the metal surface, then the metal will emit an electron with a characteristic energy $h\nu - P$.

Einstein's light-quanta hypothesis explains why red light fails to induce a photoelectric effect. Red light has a longer wavelength, and a lower frequency, then green or blue light. Because their frequency ν is smaller, the photons in red light do not have enough energy to kick out an electron. (In other words, $h\nu < P$.) If you make the light more intense while keeping the color the same, the *number* of photons will increase, but none of them are energetic enough to produce the photoelectric effect.

The strange behavior of photons has real consequences. For example, in recent years, there has been controversy over an alleged link between mobile phone usage and cancer. Holding a cell phone next to your head increases the intensity of the radiation that you are exposed to. Therefore, it seems like common sense that this will increase the risk of radiation-induced damage to your cells. However, this is completely wrong. If the general public understood Einstein's century-old discovery of the quantization of light, the mobile-phone health scare would never have gotten started.

Radiation damages biological tissues by knocking electrons loose from atoms, making the atoms more reactive. Just as in the photoelectric effect, it is the frequency of the radiation that matters, not the intensity. The infrared radiation emitted by a cell phone has even lower frequency than red light, so it is below the threshold where it can knock an electron loose from a metal atom. Moreover, our bodies are not made of metal! The atoms in our bodies hold their electrons more tightly than metals do. For both of these reasons, infrared light is completely safe to us; it cannot ionize the atoms in our body. Red, green ,and blue light are also perfectly safe. Otherwise we would have to live in caves, and avoid exposing ourselves to green grass and a blue sky.

Opposite Atomic burst over Hiroshima, from the first atomic bomb used in military action, nicknamed "Little Boy."

We need to start worrying about cancer only when we are exposed to higher frequencies of radiation, such as ultraviolet light or X-rays. If the frequency is high enough to ionize the atoms in your body, then the intensity will start to matter—but not until then.

FOR THE PHYSICISTS of Einstein's day, the light-quanta hypothesis flew in the face not only of common sense, but also of a century of theory. The debate over whether light was a particle or a wave had been going on since the early 1800s, and had apparently been resolved in favor of waves. Maxwell's equations proved that light is an electromagnetic wave.

Now Einstein was reopening a question that had seemed to be settled for good. Older physicists did not take kindly to it. In 1913, Max Planck wrote, "[Einstein] might sometimes have overshot the target in his speculations, as for example in his light quantum hypothesis." In 1916, Robert Millikan wrote that the hypothesis "may well be called reckless." Nevertheless, Millikan's own experiments showed that Einstein was correct, and six years later Einstein won the Nobel Prize for his explanation of the photoelectric effect.

In fact, Einstein's equation $E = h\nu$ explained much more than one

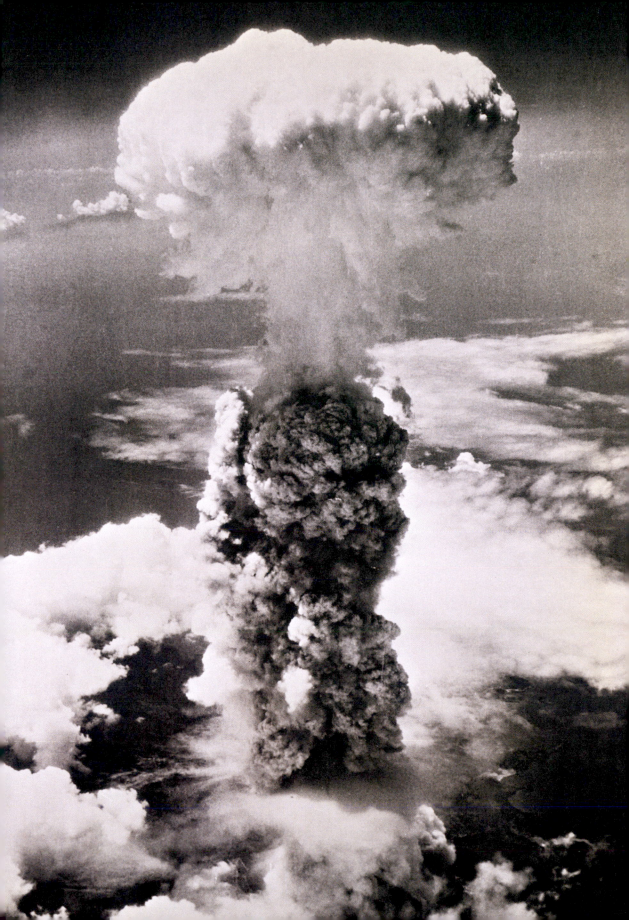

minor experimental effect. It was the first shot in the quantum revolution. Quantum mechanics resolved the age-old "particle versus wave" debate in an utterly paradoxical fashion. Light is both a particle *and* a wave. Which one it "looks like" depends on how you interrogate it. If you measure its frequency and wavelength, light looks like a wave. If you count photons by using the photoelectric effect, then light looks like a particle.

Any attempt to describe this wave-particle duality in common language tends to fall flat, because there is nothing like it in our experience of the macroscopic world. Intuition and common sense often mislead us, as the cell-phone example shows. In the subatomic world, mathematics is the only reliable guide.

If Einstein had been any ordinary scientist, or even any ordinary Nobel laureate, his equation $E = h\nu$ would have been the achievement of a lifetime. Instead, it is not even his most famous equation beginning with the letter E! That distinction, of course, falls to another equation that he first set down on paper in 1905:

$$E = mc^2$$

This equation expresses, with elegant simplicity, the equivalence between matter and energy. A particle of mass m contains an amount of energy E that is equal to the product of the mass and the square of the speed of light (c^2). Because the speed of light is a gigantic number, even a tiny amount of matter can be converted to vast amounts of energy. The bombing of Hiroshima in 1945 proved this only too well. The "Little Boy" bomb contained about 140 lb (64 kg) of enriched uranium. The amount of matter actually transformed into energy in the explosion was a little more than the weight of one BB pellet. One BB pellet was enough to destroy an entire modern city.

When Einstein discovered the equivalence of energy and mass, in 1905, he did not yet foresee its terrible consequences. He wrote to a friend: "The principle of relativity in relation to Maxwell's equations demands that mass is a direct measurement of the energy of a body; that light carries mass. A noticeable decrease in mass must then occur in the case of radium. The thought is funny and infectious; but whether God is laughing and has led me by the nose, I do not know." Forty years later, the whole world found out that God was not laughing.

How did Einstein arrive at his most famous equation? It goes back to his second great discovery of 1905: the theory of special relativity.

NEWTON'S LAWS OF MOTION hold true in any inertial frame of reference—that is, one moving at a constant velocity. If you are traveling in a train at a constant speed of 100 meters per second, it feels just the same to you as if the train were standing still and the surrounding landscape were rushing by you at 100 meters per second. No physical experiment can tell you the difference. On the train, bodies moving in a straight line will continue to move in a straight line (Newton's first law). On the train, an applied force will produce an acceleration according to the equation $F = ma$ (Newton's second law). On the other hand, if the train suddenly slows down or speeds up, so that its velocity is not constant, you can detect this fact.

However, Maxwell's equations for electromagnetism do not seem to behave the same way. Recall that the speed of light appears in Maxwell's equations as a physical constant. Therefore, if the relativity principle applies to Maxwell's equations, any measurement of the speed of light in any inertial reference frame should give the same result—299,792,458 meters per second.

But that fact leads to a paradox. If a car is traveling at 120 meters per second, and you follow it on a train traveling at 100 meters per second, then it should seem to you as if the car is going much slower—only 20 meters per second. Likewise, if you are chasing a light wave, it should appear to move 100 meters per second slower than its normal speed (that is, 299,792,358 meters per second). Or if you are approaching a light wave head on, it should appear to be moving 100 meters per second faster than usual.

In fact, we do, sort of, live on a moving train—we call it Earth. Because our orbit around the Sun takes us in different directions at different times of year, physicists reasoned that the velocity of light coming from a distant star should appear to change, depending on whether we are moving toward or away from it. Yet many experiments, including a famous one by Albert Michelson and Edward Morley in 1887, failed to discover any such changes. Einstein learned about these experiments while he was a student.

Einstein argued that the experiments had failed because there is nothing to detect. He elevated the relativity principle to the status of a postulate: the laws of physics are the same in all inertial reference frames. In particular, this means that speed of light *is* a universal constant. We do not have to abandon Maxwell's equations; nor do we have to abandon Newton's laws. We do, however, have to modify them. The common-sense subtraction of velocities, given in the example above, is incorrect and has to be replaced by a somewhat more elaborate formula. More importantly, we have to abandon our common-sense conceptions of space and time. According to Einstein, the reason we do not detect any change in the speed of light is that lengths and time intervals are *relative*. They depend on your frame of reference.

Below Albert Einstein, (1879–1955).

Imagine you are on a train speeding past a stationary Albert at 100 meters per second. As your train goes whizzing by, Albert (if he is very observant) will notice that it has gotten a little bit shorter than it was when it was standing still, and he will also notice that the watch on your wrist is running

a little bit slower. If you synchronize your watches so that they both read 10:00 the instant that you pass him in your train car, he will see his watch reach 10:01 before yours, because of the time dilation effect. But you will insist that your watch reached 10:01 first! And you are both right. By the time that the light waves from his digital watch reach you, showing the readout 10:01, your watch will already show 10:01, and vice versa. In Einstein's universe, there is no absolute measurement of time, and in fact there is no absolute concept of "before" and "after."

The only reason we do not normally perceive the shrinking of space or the dilation of time is that we normally move at very slow speeds compared with light. Thus the effects are extremely small. However, with precision instrumentation it is possible to test the predictions of relativity theory. A clock launched into orbit and then brought back to Earth really does run a few nanoseconds slow. Global positioning satellites take relativistic effects into account. Your GPS receiver compares time signals received from several different satellites, in order to determine how far away you are from them. Each of those satellites is moving rapidly with respect to you, so their clocks will be slowed down by the time dilation effect. Thanks to GPS, we are now living in the era of applied relativity.

EINSTEIN FORMULATED two different theories of relativity, as mentioned earlier. In "special relativity," which he developed in 1905, he assumed that the laws of physics are the same when viewed from any inertial reference frame (i.e., one moving at constant velocity). However, it continued to bother Einstein that accelerated reference frames (which move at non-constant velocities) were somehow different. For several years, he sought a truly general theory of relativity in which the laws of physics would be expressed in the same way regardless of the observer's frame of reference.

His key insight was that acceleration is indistinguishable from gravity. We can see this clearly in the case of astronauts orbiting Earth. We typically talk about them being in "zero-gee" (no gravity), when in fact they are still very much within Earth's gravitational field. They do not perceive the force of gravity because they are in free fall, along with their entire spacecraft. According to Einstein's theory of "general relativity," there is no observable difference between free fall in a gravitational field and constant-velocity motion in a part of space with no gravitational field.

Mathematically, general relativity is quite a bit more difficult than special relativity. You can get a general idea of this by looking at Einstein's field equations, which replace Newton's law of gravitation:

$$R_{\mu\nu} - \frac{1}{2} g_{\mu\nu} R = \frac{8\pi G}{c^4} T_{\mu\nu}.$$

The indices μ and ν refer to the four coordinates of spacetime, and each pair of indices (00, 01, 02, 03, 11, 12, 13, 22, 23, and 33) corresponds

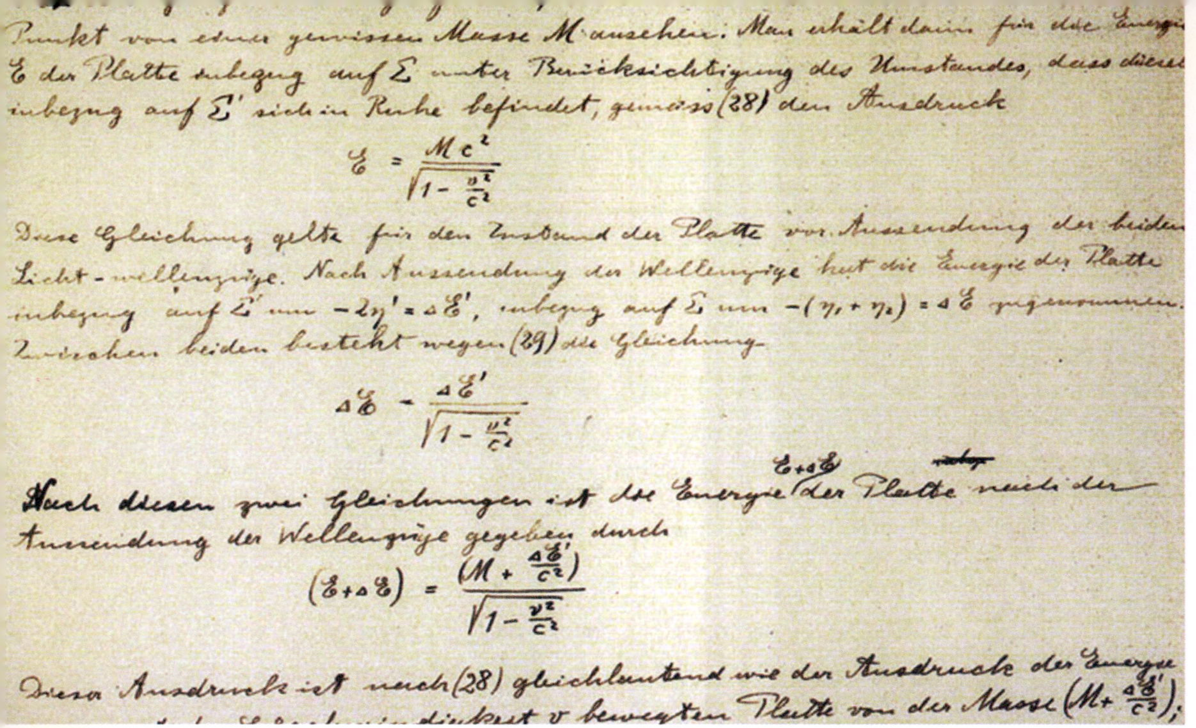

Punkt von einer gewissen Masse M ansehen. Man erhält dann für die Energie $\mathcal{E}$ der Platte inbezug auf Σ unter Berücksichtigung des Umstandes, dass diese inbezug auf Σ' sich in Ruhe befindet, gemäss (28) den Ausdruck

$$\mathcal{E} = \frac{Mc^2}{\sqrt{1-\frac{v^2}{c^2}}}$$

Diese Gleichung gilt für den Zustand der Platte vor Aussendung der beiden Licht-wellenzüge. Nach Aussendung der Wellenzüge hat die Energie der Platte inbezug auf Σ' um $-2\eta' = \Delta\mathcal{E}'$, inbezug auf Σ um $-(\eta_1+\eta_2) = \Delta\mathcal{E}$ zugenommen. Zwischen beiden besteht wegen (29) die Gleichung

$$\Delta\mathcal{E} = \frac{\Delta\mathcal{E}'}{\sqrt{1-\frac{v^2}{c^2}}}$$

Nach diesen zwei Gleichungen ist die Energie $\mathcal{E}+\Delta\mathcal{E}$ der Platte nach der Aussendung der Wellenzüge gegeben durch

$$(\mathcal{E}+\Delta\mathcal{E}) = \frac{(M + \frac{\Delta\mathcal{E}'}{c^2})}{\sqrt{1-\frac{v^2}{c^2}}}$$

Dieser Ausdruck ist nach (28) gleichlautend wie der Ausdruck der Energie ... digkeit v bewegten Platte von der Masse $(M+\frac{\Delta\mathcal{E}'}{c^2})$;

to a different equation. Thus the line above actually comprises ten separate equations.

The left side of Einstein's field equations measures the curvature of space, and the right side, the "stress-energy tensor," represents the propagation of matter and energy (which are equivalent!). John Wheeler, a leading relativity theorist, expressed the meaning of this equation succinctly: "Matter tells spacetime how to curve, and curved space tells matter how to move." The above equation led to the discovery of black holes and to the Big Bang theory, and on a more mundane level it provides additional correction terms to GPS satellites. In fact, the general relativity corrections to GPS are larger than the special relativity corrections.

It was also the general theory of relativity that led to Einstein's prediction of the curving of light rays in a gravitational field. For example, the light from distant stars bends as it passes by the Sun. When this prediction was confirmed by measurements taken during the solar eclipse of 1919, Einstein rocketed to sudden celebrity.

But now let's answer the question posed earlier: How did Einstein realize that matter and energy are equivalent? By 1915, when he wrote down the equations of general relativity, the equivalence was second nature to him. But in

Above A page from Albert Einstein's General Theory of Relativity. He donated the complete original manuscript of his ground-breaking theory to the Israeli Academy of Sciences and Humanities.

1905, when he was still just a patent clerk, general relativity was not even a gleam in his eye yet. All he had to work with was special relativity.

EINSTEIN DISCOVERED HIS MOST famous equation by pursuing a seemingly innocuous observation to its logical conclusion. He asked what would happen if a body emitted two photons in opposite directions, and if it was viewed in two different inertial frames: one at rest with the body, the other moving at velocity v, perpendicular to the photons. He showed that the photons would be blue-shifted (have higher frequency) in the moving coordinates. Thus, because of his first equation $E = h\nu$, they must also have higher energy. Einstein argued that the energy could only have come from the kinetic energy of the body that emitted them. The Newtonian formula for the kinetic energy is $1/2\ mv^2$, half the mass times the velocity squared. But the velocity of the object in the moving coordinate frame could not have changed when it emitted the photons, because their momenta cancel each other. Therefore the mass must have changed! The body has converted mass to energy, and the amount of mass converted can be computed:

$$m = \frac{E}{c^2}$$

The most famous equation in history follows as a consequence.

Einstein's paper, "Does the Inertia of a Body Depend Upon Its Energy Content?" is both beautiful and horrifying to a mathematical purist. It is only three pages long. It is wonderful to see how the strands of Einstein's thought weave together, combining the light-quanta hypothesis with special relativity like two instruments in a duet. But the "lazy dog" of a composer, the *enfant terrible* who did not care about his mathematics courses, is still very much in evidence. Einstein does not actually prove that $E = mc^2$! He makes an approximation at one point, and therefore he proves only that $E \approx mc^2$ (that is, energy is *approximately* equivalent to matter). He makes no real attempt to determine how accurate the approximation is. It's as if he couldn't be bothered. Why spoil a "funny and infectious" thought with a pedantic mathematical proof? Later, of course, Einstein and others would go back and provide more rigorous arguments for this most important physical principle.

20

from a bad cigar to westminster abbey
dirac's formula

By 1922, Albert Einstein was an international celebrity, mostly because of his theory of general relativity. Meanwhile, the quantum revolution that he had begun was continuing to progress apace, mostly without Einstein's participation. The world of physics was in turmoil, with just as many skeptics as believers in the new quantum physics. Even the believers were not sure just how much to trust their new theories.

What is quantum physics, and what is it that makes it so revolutionary? At its most basic level, it simply asserts that the measurements that physicists make, such as energy, electric charge, and angular momentum, are quantized. They are not infinitely divisible; there is a smallest unit of energy, of charge, etc.

Taken by itself, that statement may seem interesting but hardly revolutionary. The revolutionary implications become apparent when you start prying into the details. Individual quanta do not behave like anything else we are used to in the macroscopic world. For example, Einstein showed that a photon is both a particle and a wave. How is that possible? Our intuition, adapted to a universe where particles are particles and waves are waves, is helpless to explain it. At that point, mathematics becomes our only guide.

$$E\psi = \left(i\beta m + \alpha \cdot p \right)\psi$$

ψ denotes a wave function, which represents (for example) the state of an electron. E represents the electron's energy, m its mass, and p its momentum. Both α and β are spin matrices or "spinors." Dirac's equation modifies Einstein's to say that the energy of a particle depends on its mass, momentum, and spin.

Another prediction of quantum theory that seemed to require particles to perform impossible feats was the quantization of angular momentum.

At that time, it was believed that for a particle of mass m, the quantum of spin is $mh/2\pi$. (Here, h is Planck's constant, which also appeared in the formula for the energy of a quantum of light.) If you measure the angular momentum of the particle about any axis, you will get a multiple of this same quantum. Such a phenomenon would be completely impossible in classical physics. A classical particle, such as a planet or a bowling ball, has a pre-existing axis of rotation before you measure it. That axis may be askew from the direction you choose to measure. If so, you will only succeed in measuring part of the angular momentum.

But for a quantum particle, any observation of the angular momentum is an "all or nothing" proposition. Either you will see all of the angular momentum about the axis that you choose, or none of it. It is almost as if the particle waits for you to observe it, and then at that instant "decides" whether to spin around that axis or not. This "observer effect," in which the observer seems to affect the system being observed, is ubiquitous in quantum physics; remember, for instance, that a photon seems to decide whether it is a particle or a wave based on what experiment the observer chooses to perform.

Above A digital
interpretation of quantum
particles.

Two physicists in Frankfurt, Germany, saw an excellent opportunity to test, and possibly refute, the quantum theory. Otto Stern and Walther Gerlach had developed a method for producing a beam of silver atoms. When a magnetic field was applied to the beam, according to the quantum theory, the atoms in the beam would be directed right or left, depending on their axis of rotation. If they were rotating counterclockwise about the axis of the magnetic field ("spin-up") they would be deflected one way. If they were rotating clockwise ("spin-down") they would be deflected the other way. Thus the beam would split in two.

However, if the world were described by classical physics, the spin directions of the silver atoms would be randomly oriented. Some atoms would be deflected a little bit in one direction, and some would be deflected a little bit in the opposite direction, and all would be deflected by different amounts. Instead of splitting into two beams, the silver atoms would fan out into a diffuse, wider beam. Which theory would be proven right?

Unfortunately, when they first looked at their collector plate, Stern and Gerlach did not see anything! Their beam was too weak, and the number of silver atoms deposited on the plate was too small to detect.

But as Stern hunched over the plate, with Gerlach peering over his shoulder, they saw two dark lines magically appear where none were visible before. The reason, Stern later deduced, was that both he and Gerlach were cigar smokers. "My salary was too low to afford good cigars, so I smoked bad cigars," he wrote. "These had a lot of sulfur in them, so my breath on the plate turned the silver into silver sulfide, which is jet black, so easily visible. It was like developing a photographic film." (A re-enactment in 2003 showed that "cigar breath" is not strong enough to produce the effect Stern described. However, exposing the silver directly to cigar smoke does work, and presumably that is what happened.)

Thus, thanks to a cigar, the quantum prediction was confirmed: the beam split in two. However, this was not the end of the story. With hindsight, physicists now know that the effect Stern and Gerlach had observed was not what they had been looking for. Like Columbus, who went looking for India but found America instead, they had had gone looking for the orbital angular momentum of the electrons rotating about the silver atom's nucleus. What they found instead (without realizing it) was the spin of the electrons themselves. The discovery had repercussions no one could have expected.

SOME PHYSICISTS had considered the possibility that electrons could spin. However, in order to achieve an angular momentum of $mh/2\pi$, the electron would have to spin so fast that its outer surface would be traveling faster than the speed of light! Of course, according to the theory of relativity that was not possible.

A young graduate of Cambridge University, Paul Adrien Maurice Dirac, set out in 1927 to reconcile the quantum mechanics of the electron with special relativity. He started with an equation that Einstein himself had written down:

$$E^2 = m^2c^4 + p^2c^2$$

This may look somewhat familiar; it is the formula for the equivalence of matter and energy ($E = mc^2$), only it has been corrected to include the momentum of the electron (p). Another immediately obvious change is that the formula now gives the square of the energy, rather than the energy itself. Dirac was convinced this was a defect, and he looked for a way to take the square root of the equation. However, simply writing a square root in front was not acceptable to him. He had a highly aesthetic approach to physics, and many times said that the equations of physics must be beautiful. Square roots, to Dirac, were ugly.

Instead, Dirac wrote the formula for the electron in the following way:

$$E = \alpha_1 p_1 + \alpha_2 p_2 + \alpha_3 p_3 + i\beta m$$

where p_1, p_2, p_3 represent the three components of the electron's momentum in three-space. The mysterious quantities $\alpha_1, \alpha_2, \alpha_3$ and β satisfy the following relations:

$$\alpha_1^2 = \alpha_2^2 = \alpha_3^2 = -\beta^2 = 1$$

$$\alpha_1\alpha_2 = -\alpha_2\alpha_1,$$

$$\alpha_2\alpha_3 = -\alpha_3\alpha_2,$$

and $\alpha_3\alpha_1 = -\alpha_1\alpha_3$.

I have displayed these formulas to make a point: they are virtually identical to the quaternion formulas that William Rowan Hamilton had written 80 years earlier! Only the names have changed (and −1 has been changed to 1 in the first equation). Dirac had, in a sense, rediscovered quaternions, although he wrote them as 4-by-4 matrices.

The other change that Dirac made to Einstein's formula was to rewrite the energy as an operator on a wave function, ψ. This is consistent with the philosophy of quantum mechanics: any observable quantity of a particle is not merely a number but an actual physical operation applied to that particle. (This is why the observer is such an intrinsic part of quantum mechanics.)

Thus, the final form of Dirac's equation looks like this:

$$E\psi = (i\beta m + \alpha \cdot \bar{p})\psi.$$

Here I have, for convenience, condensed the three alpha matrices into one symbol α, and written the momentum as a single vector p. In order for the formula to make mathematical sense, the wave function ψ has to be a quaternion-like object with four components. This was actually the most puzzling aspect of the equation to physicists, for two reasons.

Firstly, it had four components instead of two. Physicists could make sense of two components—they would represent the spin-up and spin-down states of an electron. But what was the meaning of the other two?

Secondly, the wave function did not behave like a vector (an "arrow" in spacetime). When you rotate space by 360 degrees, the wave function rotates by only 180 degrees, and thus the electron goes from "spin-up" to "spin-down."

The second point shows that electrons are not like bowling balls or planets. However, there is an ingenious analogy that goes by the name of

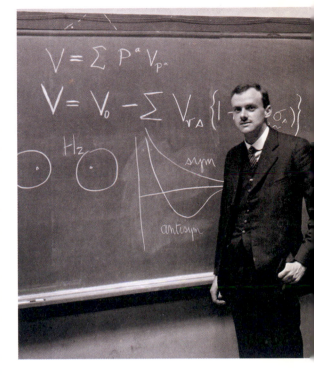

Above Paul Dirac standing in front of a blackboard displaying a quantum mechanical model of the hydrogen molecule.

the "Feynman plate trick" or "Dirac belt trick." Place a plate in your open palm in front of you. Now rotate the plate 360 degrees, by rotating your arm in a circle while keeping your palm up. You will find your arm is in quite an awkward position; unlike the plate, it has not come back to its original state. It has rotated 180 degrees. But if you continue and rotate the plate one more time, your arm will come back to its normal, comfortable position! The whole system of "arm plus plate" behaves like a quaternion.

Although Einstein's formula $E = mc^2$ may be better known to the public, Dirac's formula may well be of greater significance both to physicists and mathematicians. "Of all the equations of physics, perhaps the most 'magical' is the Dirac equation," wrote Frank Wilczek of MIT in 2002, on the centennial anniversary of Dirac's birth. "It is the most freely invented, the least conditioned by experiment, the one with the strangest and most startling consequences … [It] became the fulcrum on which fundamental physics pivoted."

WHY DID IT CHANGE physics so much? Let's start with those two extra components of the electron wave function. Dirac explained them as particles with negative energy, or "holes" in space. They should appear to be particles just like electrons, but with a positive charge. He proposed the idea in 1931, with great hesitancy. Other physicists ridiculed the idea. Wolfgang Heisenberg wrote, "The saddest chapter of modern physics is and remains the Dirac theory."

Yet within a year, Carl Anderson of Caltech had discovered Dirac's positively-charged electron, or positron, in an experiment. It was the first time that a theoretical physicist had successfully predicted the existence of a previously unknown particle for purely mathematical reasons. Nowadays, theoretical physicists do this with gleeful abandon, and they are occasionally right. Dirac's discovery utterly changed the rules of the game; the theoreticians no longer had to wait for experiments.

Opposite Electromagnetic particle shower. Particle tracks (moving from bottom to top) showing multiple electron-positron pairs created from the energy of a high-energy gamma ray photon.

The positron was also the first antimatter particle to be discovered. Physicists now understand that every particle has an antimatter equivalent; if a particle meets its antimatter twin, the two are annihilated. Thus Dirac's formula led to a new and still unsolved problem: Why do we have more matter in the universe than antimatter? Why isn't the universe empty?

Dirac's equation also revealed that our universe has two fundamentally different kinds of quantum particle. Some particles have spin 0, ±1, ±2, etc., have vector wave functions, and are known as bosons. For example, photons fit into this category. Others, such as electrons, have spin $\pm^1/_2$, $\pm^3/_2$, etc., have

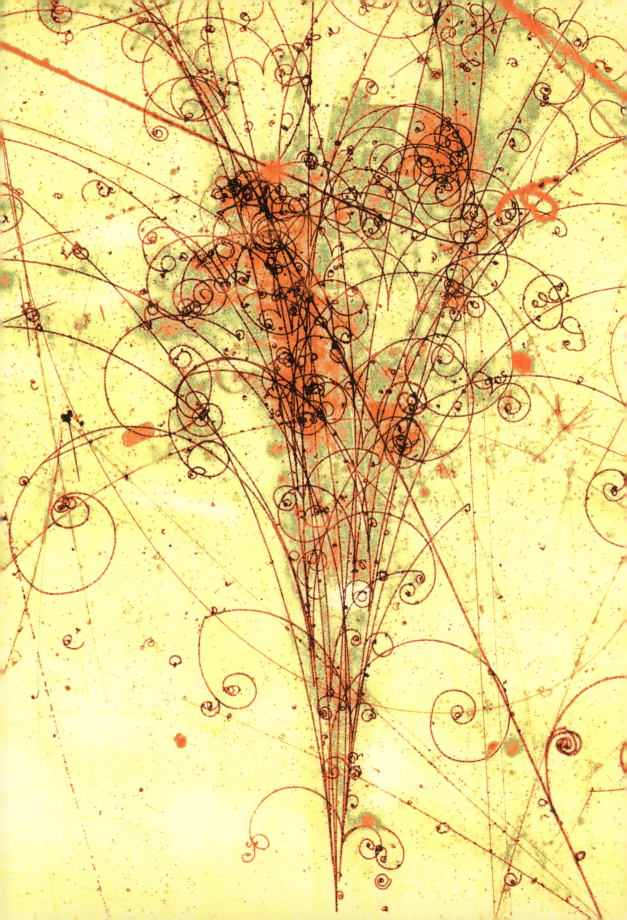

quaternion-like (or "spinor") wave functions,[*] and are known as fermions. All of the basic particles of ordinary matter—electrons, protons, and neutrons—are fermions.

Bosons like to congregate together; that is why lasers are possible. A laser beam is a collection of photons in the same quantum state. Fermions, on the other hand, stay aloof—you will never find two of them in the same quantum state. This is a good thing: it explains why atoms have electron orbitals. Because electrons can't overlap, there is only room for two of them at the lowest energy level of an atom, eight at the next energy level, and so on.

This pattern explains the periodic table and underlies all of chemistry. Imagine a universe without Dirac's equation: it would be a universe with no matter as we know it, no chemical reactions, a universe with light and nothing else. A universe frozen at the first sentence of Genesis!

NOW LET'S STEP DOWN from the mountaintop and look at the more mundane applications of Dirac's equation. They are legion. I have already mentioned lasers. Also, positrons are the fundamental ingredient in positron emission tomography (PET scans), used to study the activity of the brain. Electron spins are manipulated by magnetic fields in magnetic resonance imaging (MRI scans), a tool used to diagnose diseases without exposing patients to X-rays.

Finally, Dirac's equation led quantum physicists to a new understanding of the vacuum, the ground state of the universe. They no longer see the vacuum as empty, but teeming with energy. Particles and their antiparticles can, and do, routinely pop into existence and pop right back out again. In fact, the whole concept of a "particle" is slightly outdated. To quantum physicists, the

---

[*] Experts may object to my identification of spinors with quaternions. In fact, the difference of a minus sign in their definitions has an important consequence: spinors exist in all dimensions, while quaternions are to some extent a low-dimensional fluke. However, in three and four dimensions the quibble is purely academic. Three-dimensional spinors are quaternions of length one, or "unit" quaternions. Four-dimensional spinors are pairs of unit quaternions.

really fundamental concept is a *quantum field*. These fields, like electric fields, pervade all of space, and particles are their local manifestation. A particle is a fluctuation in the quantum field that may be just temporary or may be long-lasting.

Few equations in history have had more far-reaching implications. However, the man who discovered it, Paul Dirac, was notoriously taciturn and shy of publicity. If he spoke two words in a conversation, it meant that he was in a talkative mood. When he found out that he was going to receive the Nobel Prize in 1933, he initially wanted to decline it, until his friends persuaded him that declining the award would cause him to receive more publicity than accepting. Dirac largely escaped the public fascination and adulation that followed Newton and Einstein.

Nevertheless, Dirac was certainly appreciated by his colleagues. He inherited Newton's professorship at Cambridge (the Lucasian Chair), and after he died in 1984, a memorial plaque in his name was placed in Westminster Abbey, not far from Newton's grave. Appropriately for the man of few words, the plaque includes his equation. It is the only formula that has been preserved for posterity in the church.

21

the empire-builder
the chern-gauss-bonnet equation

Although it is impossible to summarize a whole century of mathematics in a few sentences, or even in a few pages, some trends can be discerned. The connection between physics and mathematics, which had always been close, in the twentieth century became deeper and more mysterious. Physicists, beginning with Einstein, were routinely startled to discover that mathematicians had already developed the tools they needed. Vice versa, mathematicians kept on discovering that the problems and equations of physics led to the most interesting and deepest mathematics.

Another trend in the twentieth century, connected with the first, was the rise of geometry. Einstein's theory of general relativity required space to be curved, and this demanded a non-Euclidean geometry whose curvature could vary from point to point. Gauss, Lobachevsky, and Bolyai had sowed the seeds of non-Euclidean geometry in the early 1800s, but their geometries had constant curvature. Bernhard Riemann made it possible to vary the curvature. Riemannian, or differential, geometry developed rather slowly for the first half of the twentieth century, but in the second half it exploded and became a central area of mathematics.

A third important, and invigorating, trend in mathematics was its increasing globalization, especially after the Second World War. Many

$$\int_{M} Pf(\Omega) = (2\pi)^n \chi(M)$$

M represents an even-dimensional space or universe with no boundary. $\chi(M)$ is the Euler characteristic of the space, which in two dimensions tells you the number of holes it has. Ω is the curvature of the space. The formula allows you to deduce information about the overall shape of the universe if you know its curvature at every point.

parts of ancient mathematics had been discovered in Asia, Egypt, or the Arab world before they reached western Europe. But from roughly 1500 to 1900, mathematics was mostly a game for European males. Now, with the inequality of opportunity decreasing, the next great discovery is now (almost) as likely to come from a Zhang or an Alice as from a Smith or a Bob.

If I could choose one figure to exemplify all three of these trends, it would be Shiing-Shen Chern. Born in Jiaxing, China (near Shanghai) in 1911, Chern attended Nankai University near Beijing. He distinguished himself enough there to earn a scholarship to study in Europe. He studied for two years in Frankfurt with a geometer named Wilhelm Blaschke, then moved to Paris for a year to work with Elie Cartan.

At that time, differential geometry was not a very fashionable subject. Looking at Einstein's field equations on page 161 may give you a sense of why. In order to describe the geometry of a curved space (or "manifold"), you need to establish a set of coordinates on it. When equations are written in terms of these coordinates, they are festooned with symbols (like the indices μ and ν in the field equations) that act merely as bookkeeping devices. Michael Spivak, author of a classic textbook on differential geometry, calls it "the debauch of indices."

Ironically, the most important and interesting quantities in differential geometry are precisely those that do not depend on the choice of coordinates. In other words, we spend all this time keeping track of something that in the end we don't care about! For example, in Einstein's theory of general relativity, the independence of physical laws from the coordinate system was a fundamental tenet. Yet it took Einstein years to navigate the mathematics and find equations with the appropriate invariance.

Cartan, Chern's mentor, had pioneered an approach to differential geometry, called "moving frames," which worked without coordinates. However, Cartan's theory was extremely obscure and difficult to understand. Chern became his foremost interpreter for the rest of the world. In the process he transformed Cartan's theory from a local one, suitable for describing small pieces of a curved space, into a global one that dealt with space as a whole. Chern's first famous result, considered by many (including Chern himself) to be his greatest work, was a generalization of a nineteenth-century theorem about surfaces that had been named after Karl Freidrich Gauss and Pierre Ossian Bonnet. The Chern–Gauss–Bonnet theorem, as it is now known, reads as follows:

$$\int_M Pf(\Omega) = (2\pi)^n \chi(M).$$

What does this mean? From a top-level view, it means that if we live in a curved space or "manifold" (here denoted M), we can learn something about the global shape of our universe (here denoted $\chi(M)$, the Euler characteristic of M) by measuring the curvature (Ω) at every point. The *Pfaffian* (Pf) is a specific computation we must do with the curvature, and the integral sign ($\int$) means that we have to add up the curvatures of every point in the manifold. This is a global theorem *par excellence*.

Let's burrow a little bit deeper. In this formula the curved space M is assumed to be ($2n$)-dimensional. So in the simplest case, where $n = 1$, we are dealing with a two-dimensional space, or a surface. Surfaces have only one local geometric property that is independent of the coordinate frame, called the Gaussian curvature, K. If the surface is convex the Gaussian curvature is positive. If it is shaped like a potato chip, the curvature is negative. The total curvature within any region on the surface is a measure of how much the geometry differs from Euclidean. For instance, if the total curvature within a triangle is x, then the sum of the angles of the triangle will be:

$$\left(180 + \frac{180x}{\pi}\right) \text{ degrees}$$

If the curvature is zero, the sum of the angles is 180 degrees, as Euclid had shown. For example, on a sphere, which is positively curved, you can find a triangle with three right angles. (See my discussion of Ant Geometry in Part Three, page 161.) The sum of the angles of this triangle is 90 + 90 + 90 = 270 degrees, and hence the total curvature inside it must be $\pi/2$.

Let's check this prediction against the Gauss–Bonnet formula. The curvature at every point on the sphere is

$$\frac{1}{R^2}, \text{ where } R \text{ is the radius of the sphere.}$$

The total curvature within the triangle is obtained by multiplying this curvature by the area of the triangle. The area of a sphere is $4\pi R^2$, and it takes eight right triangles to cover the sphere. Thus each triangle has an area of:

$$\frac{4\pi R^2}{8} \text{ or } \frac{\pi R^2}{2}$$

Multiplying the area by the curvature gives a total curvature of:

$$\left(\frac{\pi R^2}{2}\right)\left(\frac{1}{R^2}\right) \text{ or } \pi/2, \text{ as promised.}$$

THE GAUSS-BONNET THEOREM demarcates a transition point between ancient geometry (What is the sum of the angles of a triangle?) and modern geometry (How do we describe the global properties of a curved surface?). From now on, we will leave ancient geometry behind. In order to make the Gauss–Bonnet theorem global, we need to add up the curvature not only over one triangle, but over our whole surface.

When we do this, we make an extraordinary discovery. If the surface M is roughly ball-shaped—it can be a sphere, a football, or anything else without a hole—then its total curvature will always come out to 4π. That is:

$$\int_M K = 4\pi.$$

If we do the same computation on any surface M that is more or less torus-shaped—it can be a doughnut, a coffee cup, a vuvuzela, or anything else with one hole—then the total curvature is 0. That is,

$$\int_M K = 0$$

More generally, if the surface has g holes in it, then the total curvature allows us to detect the number of holes:

$$\int_M K = 2\pi(2 - 2g) = 2\pi\chi(M).$$

The number $\chi(M) = 2 - 2g$ is the Euler characteristic of the surface. This formula matches the Chern–Gauss–Bonnet formula given above (with $n = 1$, of course).

The classical Gauss–Bonnet theorem for surfaces is remarkable for two reasons. First, it means that a very smart ant, using only Ant Geometry, can determine what kind of surface it is crawling on (a ball, a torus, or something more complicated). The curvature K is intrinsic to the surface, which means that an ant does not have to go outside the surface to measure it.

Second, the total curvature of a surface is *quantized*. It is always a whole number times 2π. Thus this nineteenth-century formula foreshadows the preoccupation of twentieth-century mathematicians and physicists with quantization.

IT IS NOW TIME to call Shiing-Shen Chern back from the wings. In 1943, Chern was extricated from Japanese-occupied China by the US Army and went to the Institute for Advanced Study in Princeton. While he was there, he heard that two other mathematicians, Andre Weil and Carl Allendoerfer, had proved a version of the Gauss–Bonnet theorem that worked for any even-dimensional curved space or manifold, not just two-

Above An artistic digital interpretation of a torus-shaped universe.

dimensional surfaces. However, their proof was ugly and unenlightening. It made an extra assumption that was later proved to be unnecessary. It resembled the Disney story of *Dumbo* who learns to fly with the aid of a magic feather and then finds out that the feather was never magic, and he could fly all along.

In a short, six-page manuscript published in 1946, Chern gave a proof that had none of these flaws, and that set postwar geometry on a new course. He introduced a concept called a fiber bundle, which is like a castle that has the manifold M as its floor plan. Everything that happens in the manifold is merely a pale reflection of what happens in the fiber bundle above it. In particular, Chern discovered that the curvature integrand (Ω) lies at the base of a tower of similar integrands, called differential forms, in the fiber bundle. When the curvature is integrated "upstairs" in the fiber bundle, rather than downstairs in the manifold, the Chern–Gauss–Bonnet theorem becomes almost obvious.

Almost, but not quite. Chern's calculation was a *tour de force*, and his idea of doing the integral in the fiber bundle was a stroke of genius. The proof made it evident that fiber bundles contained an untapped wealth of information about a space. Not only the Euler characteristic, but also a variety of other invariants, now called Chern characteristics and Chern–Simons invariants, have now been constructed in this way.

Chern's work completed a cycle. Einstein and Dirac had shown that you could not do physics without geometry. Chern showed that you cannot do geometry without physics. The fiber bundle is the building in which a quantum field lives. To understand the shape of a space, you need to know what kinds of fiber bundles—or, what is essentially the same thing, what kinds of quantum fields—can be erected on that space.

TWO DECADES LATER, in 1963, Michael Atiyah and Isadore Singer made the links between math and physics even more explicit. They gave a proof of the Chern–Gauss–Bonnet theorem (and quite a bit more) that proceeds directly from solutions to the Dirac equation!

Why does the Chern–Gauss–Bonnet theorem matter? Because if we ever want to understand the kind of universe we live in, we do not have the option of going "outside" the universe. We will have to work from within, using the language that Chern pioneered.

At the same time, I want to emphasize that mathematics is not only about the universe that we live in. To me, that is one of the main distinctions between mathematics and physics. Physics is supposed to be about our universe, and physical theories eventually have to be grounded at some point in experiment. On the other hand, mathematics is about all possible universes, the one we live in and those we do not. It is an amazing fact that to understand any possible universe (or at least any universe that is a smooth manifold with an even number of dimensions) you need the same language of fields and the same Dirac equation. For readers who are inclined to believe in a Creator, he (or she or it) must have been a very good mathematician!

Chern's later career straddled two continents. He returned to China after the war, but was forced to leave again before the Communist government took power in 1949. He enjoyed a long and successful career in the United States, first at the University of Chicago and then at the University of California at Berkeley. Together with Singer and Calvin Moore, he founded the Mathematical Sciences Research Institute in Berkeley, the first pure-math institute in the United States, and served as its first director. After he retired from MSRI in 1984, he devoted himself to reviving Chinese mathematics, which had suffered greatly during the Cultural Revolution. He traveled to China frequently and obtained opportunities for Chinese graduate students to study in America. He also founded the Nankai Institute of Mathematics in Tianjin (which was renamed the Chern Institute after his death in 2004). Remarkably, he achieved the same sort of rock-star celebrity in China that Einstein did in the United States.

Robert Bryant, a later director of MSRI, tells the story of how Chern went to watch the world table tennis championships in Tianjin in 1994. "The TV cameras were all there, showing the prime minister as he was seated," says Bryant. "Then Mr. and Mrs. Chern walked in. The cameras went straight to them and ignored the prime minister! He was this iconic figure, a great intellectual figure who had showed what Chinese could do in the outside world."

Like Dirac, Chern was very modest about his accomplishments, but unlike Dirac he was comfortable with people. He understood that mathematics advances not only by deriving formulas, but also by building institutions, such as MSRI and the Nankai Institute. According to Hung-Hsi Wu, a long-time colleague of his at Berkeley, "He was an empire-builder in the best sense of the word."

22

a little bit infinite
the continuum hypothesis

Beginning in the 1870s, mathematicians began to realize that infinity actually comes in different sizes; a set can in fact be a little bit infinite or a whole lot infinite. The exploration of these different kinds of infinity has led to some of the most profound and paradoxical discoveries of twentieth-century mathematics.

For most of the nineteenth century, mathematicians did their best to finesse the whole issue of infinity. There was good reason for that; as we have seen in Part One, the notion of infinity had been confounding mathematicians at least since the days of Zeno. In 1831, Gauss expressed this interdiction in a letter to Heinrich Schumacher: "I must protest most vehemently against your use of the infinite as something consummated, as this is never permitted in mathematics. The infinite is but a figure of speech ..."

However, toward the end of the century, a consensus began to form that sets, rather than numbers, are the fundamental building blocks of mathematics. And you just can't get around the fact that some sets are infinite: for example, the set of positive integers, {1, 2, 3, ...}, or the set of numbers that form successive approximations to pi {3.1, 3.14, 3.141, 3.1415, ...}.

Great scientists, like Einstein, are often the ones who are willing to face the inconvenient facts that other scientists would prefer to avoid. In the case

$$2^{\aleph_0} = \aleph_1$$

$\aleph_0$ is the "size" or cardinality of the smallest infinite set (the integers). $\aleph_1$ is the cardinality of the next smallest. If true, this formula would mean that the real numbers are the next smallest set after the integers.

of set theory, it was another German mathematician named Georg Cantor who blazed a trail into the strange world of infinity.

To understand Cantor, we first need a way to describe what we mean by the "size," or cardinality, of a set. First, Cantor suggested the following rule, a comparative or relative approach to defining the meaning of cardinality: If we can find a one-to-one matching between two sets A and B, so that each element of A corresponds to a unique element of B and vice versa, then the two sets have the same cardinality.

A good example (suggested by David Hilbert) is to think of A as the set of rooms in a hotel and B as the set of guests wanting a room. Ideally, we would like to place every guest in a separate room. In addition, if we are the proprietors of the hotel, we would like to fill up all the rooms. If we can do this, then the "number of guests" is the same as the "number of rooms."

Let's suppose that the hotel has infinitely many rooms, which are labeled 1, 2, 3, and so on. If a set of guests arrives that can fill up the hotel without overfilling it, the set is countably infinite. With this preamble, we can list the surprising facts that Cantor discovered about countably infinite sets:

1. If you add one element to a countably infinite set, you get a set of the same size! For instance, suppose you have filled your hotel, and then one

more guest arrives. You do not have to turn him away! You simply bump the guest in room 1 to room 2, the guest in room 2 to room 3, and so on. Presto, room 1 is now available for your new guest.

2. If you combine two countably infinite sets, you get another set of the same size. Suppose the hotel is filled with a countably infinite number of guests, but then a second countably infinite party arrives. No problem! You can accommodate them too. Just bump the guest in room 1 to room 2, the guest in room 2 to room 4, the guest in room 3 to room 6, and so on. Now rooms 1, 3, 5, etc., are free for your infinite party of new guests.

3. Proceeding in similar fashion, the union of a countably infinite number of countably infinite sets is still countably infinite. This is a little bit tricky to explain in words, but the idea is shown visually below.

All of these facts are different from our everyday experience with finite sets, and frankly they seem a little bit like magic. But perhaps this should not surprise us too much. Physicists had to leave common sense behind when they encountered quanta, and infinite sets are also a strange new world.

HAVING EXPLAINED what it means to say that two sets have the same size, Cantor next invented an absolute measure of size, called cardinal numbers. Again, finite sets are easy. A set with one element has cardinality 1, a set with two elements has cardinality 2, and so on. Cantor proposed a new cardinal number to denote countably infinite sets: aleph-nought, or $\aleph_0$.

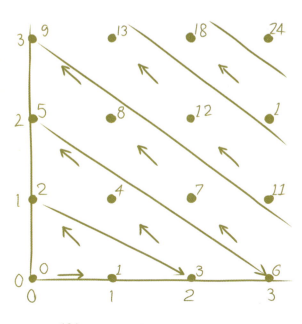

Left A countably infinite collection of countable sets (here shown as an infinite array) can still be counted. The arrows show how to place the array into a linear sequence.

So far, I have yet to show you any set with cardinality greater than $\aleph_0$. In fact, the three examples above show that $\aleph_0 + 1 = \aleph_0$, $\aleph_0 + \aleph_0 = \aleph_0$, and $\aleph_0 \cdot \aleph_0 = \aleph_0$. However, Cantor showed that the set of all real numbers does have greater cardinality; that is, the real numbers cannot be accommodated in our countable hotel. The proof of this fact, called Cantor's diagonal argument, is one of the most original and fundamental breakthroughs of modern mathematics, yet short enough to explain in a page.

Suppose that I try to come up with a room assignment for all the real numbers between 0 and 1. For the sake of argument, let's say that the list starts as follows:

Room 1: 0.**1**415926 ... (the decimal part of pi)

Room 2: 0.7**1**82818 ... (the decimal part of Euler's number *e*)

Room 3: 0.41**4**2135 ... (the decimal part of √2)

Room 4: 0.500**0**000 ...

Room 5: 0.1011**0**01 ... (I'm running out of interesting numbers, so this is just a random string of 1's and 0's).

Notice that the first digit of the first number is highlighted, as is the second digit of the second number, and so forth. Now Georg Cantor comes along and asks me, "Which room is this number in?" and he writes down the number 0.**22511** ...

It's no secret how Cantor got this number: he simply took each of the bold numerals and added 1 to it. (If he had encountered the numeral 9, he would have changed it to a 0.) Cantor's number cannot be in room 1 because its first digit disagrees with the number there. It cannot be in room 2 because its second digit disagrees with the number there. In fact, for the same reason it cannot be in any of the rooms in the hotel. The room assignment is incomplete! More importantly, Georg could repeat this procedure for *any* room assignment I came up with. Thus there is no way to accommodate the real numbers between 0 and 1 in the hotel (and so, of course, there is no way to accommodate all the real numbers).†

---

† This argument has a slight flaw in it due to the fact that some numbers have two decimal representations, for example 0.499... = 0.500... I have intentionally given this easier but flawed version for the non-experts. For math experts, repairing the inaccuracy takes a little work but in my opinion no fundamentally new ideas.

To make intuitive sense of Cantor's diagonal argument, I like to think of the number line as consisting of a countable set of bricks (the rational numbers) with a sort of "glue" of irrational, transcendental, and mostly just plain random numbers filling the gaps between them. Cantor's argument shows that the vast majority of the number line is made of glue, not bricks.

Having discovered a cardinal number greater than $\aleph_0$, we need a name for it. The customary name, which emphasizes this bricks-and-mortar intuition about the number line, is the cardinality of the continuum, or c for short. The "continuum" is all of that glue that is hard to get off your hands.

Cantor also discovered a second, and more general, way to produce higher cardinalities. If S is a set, the set consisting of all of its subsets is called the *power set* of S. Let's look at a few examples.

If S is a set with one element, say {1}, then it has two subsets, namely the empty set $\varnothing$ and the entire set {1}. Thus the power set has two elements.

If S is a set with two elements, say {1, 2}, then it has four subsets, namely $\varnothing$, the set {1}, the set {2}, and the set {1, 2}. Thus its power set has four elements. Note that $4 = 2^2$.

Similarly, if S is a set with three elements, I invite the reader to check that the figure shown here has eight subsets. Note that $8 = 2^3$.

At this point the pattern seems clear: the power set of S always has more elements than S. In fact, if the cardinality of S is n, then the cardinality of the power set is always 2^n.

With a slight modification of the diagonal argument, Cantor showed that the same thing

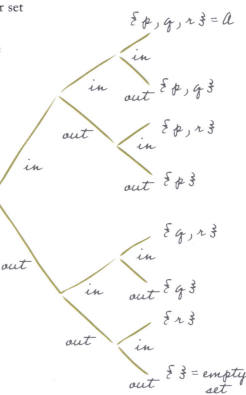

is true for infinite sets: the power set of S always has greater cardinality than S itself. So there is no largest infinite cardinal number. The infinite cardinals form a vast tower that we cannot even begin to comprehend.

If we cannot ever scale the upper reaches of this tower, perhaps we can at least comprehend the beginning. We know that the "smallest" infinite cardinal is $\aleph_0$. We know that the continuum, c, has a bigger cardinality, although we're not quite sure yet just how much bigger. The power set of $\aleph_0$ also has larger cardinality than $\aleph_0$, which we can denote $2^{\aleph_0}$ (by analogy with the pattern for finite sets).

Now Cantor asked: What is the next cardinal number after $\aleph_0$? Is it c? Is it $2^{\aleph_0}$? He partially answered this question, demonstrating that $c = 2^{\aleph_0}$. In other words, the continuum and the power set of the integers have the same size. But could there be a cardinality that is *intermediate* in size between $\aleph_0$ and $2^{\aleph_0}$—a set that is "a little more pregnant" than the integers but "a little less pregnant" than the real numbers? Cantor believed the answer was no. In other words, the next cardinal number after $\aleph_0$, which we denote by $\aleph_1$, is $2^{\aleph_0}$. Because this is a book about equations, here is Cantor's Continuum Hypothesis in equation form: $2^{\aleph_0} = \aleph_1$.

CANTOR DID NOT LIVE to see his question answered. He spent the last few years of his life confined to a mental institution, and died in 1918. Though it is tempting and facile to suggest that the lack of acceptance for his ideas "drove him crazy," it is a temptation that must be resisted. Depression is too complex an illness to reduce to such a simple formula. What is true is that his work was deeply controversial. Some mathematicians of his era, such as Leopold Kronecker, rejected it in scathing terms, while others, such as David Hilbert, strongly endorsed it. "No one shall expel us from the paradise that Cantor has created," Hilbert wrote.

In 1900, when Hilbert compiled a list of the 23 most important problems for mathematicians to work on in the twentieth century, he placed the Continuum Hypothesis at the top of the list. As it turned out, the solution of the Continuum Hypothesis would be inextricably linked to the solution of the second problem on Hilbert's list: a proof of the consistency of mathematics.

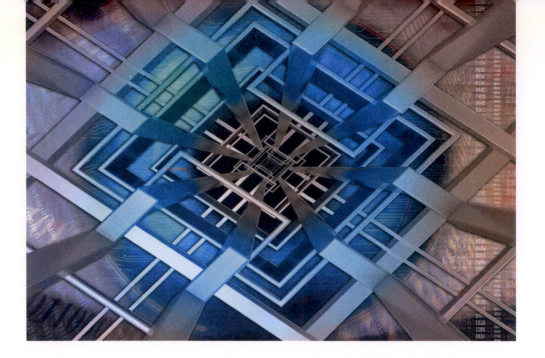

Above Conceptual artwork of a metal structure conveying the idea of an infinite depth.

By the early years of the twentieth century, the foundations of mathematics were in crisis. This seems to happen periodically in math history. The discovery of irrational numbers in ancient Greece provoked a crisis that led the Greeks to turn toward geometry instead of arithmetic as the foundation of mathematics. The discovery of calculus produced another crisis, because of its apparent manipulations of infinite and infinitesimal quantities. In reaction, mathematicians rejected "completed infinities," as the quote from Gauss earlier in this chapter attests, and reformulated calculus to use only "potential infinities."

Cantor's work on infinite sets set off another crisis. First, Cantor himself realized that there was no such thing as the set of all sets. If such a set existed, it would perforce have the largest cardinality possible, and yet we have just said that there is no largest cardinality. In 1901, the philosopher Bertrand Russell produced an even simpler paradox. Let R be the set of all sets that do not contain themselves. Does R contain itself?

Well, let's suppose it does. Then it is a set that contains itself, and therefore it is not an element of R, because R only contains sets that do *not* contain themselves. But that means R is not an element of R—contradicting the assumption that we just made that R contains itself!

These examples showed that mathematicians do not have complete freedom to create any set that we can describe in words. "Naïve set theory" does not work. There must be some rules governing what is and is not permissible in set theory.

In 1908, Ernst Zermelo drew up a list of seven axioms for set theory that ruled out Cantor's and Russell's paradoxes. With some modifications in the 1920s by Abraham Fraenkel (which brought the number of axioms to nine), his system has become the standard working foundation for the vast majority of practicing mathematicians.

In light of these developments, Hilbert's second problem was reinterpreted as follows: prove that Zermelo–Fraenkel (ZF) set theory is consistent. If this could be done, then mathematicians could sleep soundly at night, knowing that there would be no more crises in the foundations of mathematics from paradoxes we have not thought of yet.

By 1928, Hilbert was optimistic that his problem was near to being solved. At that year's International Congress of Mathematicians, he announced that only a few more details needed to be put into place in the proof. But within less than two years, the whole enterprise unraveled completely.

WITH HINDSIGHT, Hilbert's mathematical epistemology, called formalism, seems like a strange one. Hilbert believed that mathematics was essentially a formal game played with symbols. The symbols themselves have no intrinsic meaning. Every statement made with these symbols should be either true or false, and if true it should be provable from the axioms. (In other words, mathematics is *complete*.) Finally, it should never be possible to prove a statement to be both true and false; mathematics must be *consistent*.

It is a good thing that Hilbert's vision turned out to be flawed, because he was proposing to rescue mathematics by killing it. If mathematics loses its content, then it also (I believe) loses most of its beauty. If mathematicians truly believed that they were doing nothing but pushing symbols around, I think that many of them would want to find another way of passing the time.

The person who rescued mathematics from Hilbert's poorly conceived lifebuoy was Kurt Gödel, a young Austrian logician who was born in Brno (now part of the Czech Republic) in 1906. In 1930, Gödel proved that in any axiomatic system that is strong enough to include the normal rules of arithmetic, there must be statements that are both *true* and *unprovable*. In other words, mathematics is incomplete. Very shortly after his first theorem, Gödel proved a second, more specific version: the consistency of the axiomatic system itself is unprovable.

HOW CAN YOU POSSIBLY prove such a theorem as the incompleteness of mathematics? Gödel's proof contains two main ideas. The first one is an adaptation of a very ancient paradox. The Liar's Paradox concerns the following statement: "This statement is false." If the statement is true, then it proclaims its own falsity. We are stuck in the same type of self-referential vicious circle that we were with Russell's Paradox.

Below Portrait of the Austrian-US logician and mathematician Kurt Gödel (1906–1978).

Gödel's version of the Liar's Paradox reads as follows: "This statement is unprovable." If Gödel's sentence is true, then it cannot be proved. If the statement is false, then it is both false *and* provable, which means that our axiom system is inconsistent. If we assume the consistency of ZF set theory as a premise, then the second possibility is ruled out, and therefore Gödel's sentence is true and unprovable.

So far, though, we have not done any mathematics. The second, and even more ingenious, part of Gödel's proof is a way of converting the non-mathematical statement "This statement is unprovable" into a mathematical statement about numbers. In any axiomatic system, Gödel realized, there are only $\aleph_0$ possible statements. That is true because we have only a finite alphabet of possible symbols, and only a countable list of possible statement lengths. Therefore every statement—true, false, provable, unprovable, or just plain nonsensical—can be assigned a unique identifying number. Let's imagine now that we have a list L of all the identification numbers of provable statements.

Now the assertion that a particular statement is unprovable reduces to a mathematical assertion: "This number is not in the set L." Gödel uses a version of Cantor's diagonal argument to show that at least one sentence of the form "This number is not in the set L" is, in fact, *not* on the list of provable statements. Thus that particular statement is both true and unprovable.

Gödel's proof takes some getting used to; it is easy to get lost in the multiple layers of self-reference. It is intricately linked to Cantor's diagonal argument and to the counterintuitive nature of infinite cardinals. You would expect that there are many more statements about the integers than there are integers themselves. But that is not true. The fact that you can encode every statement about the integers *as* an integer, and thus transform a meta-mathematical statement into a mathematical statement, is sheer magic.

I like to think of Gödel's Incompleteness Theorem as a counterpart to the Heisenberg Uncertainty Principle. (This is the statement from quantum physics that you cannot measure both the position and the momentum of a particle at the same time; more generally, that the act of measuring certain quantitites will alter the system in a way that destroys information about other quantities.) Both of them circumscribe the limits of what humans can possibly know—one in the domain of mathematics, the other in the domain of physics. Both results were discovered within ten years of each other. After a nineteenth century that was dominated by the Victorian belief in progress and the perfectibility of human knowledge, the twentieth century was an era when humans began to become aware of their own limitations. These two landmark discoveries are a part of the philosophical Gestalt of an era.

Though the philosophical implications of Gödel's theorem were huge, its mathematical implications were curiously muted. In some ways, mathematicians shrugged their shoulders and went on with their business. From today's perspective, the theorem is similar to the asteroid that killed off the dinosaurs. It killed off some mistaken ideas about mathematics, and it left a large crater, but today that crater is virtually impossible to detect.

Why is its influence so undetectable? One reason, I think, is that Gödel's unprovable statements are very artificial. The statement "mathematics is consistent" can be written down in words, but its mathematical version, after it has been converted to a statement about numbers, is impossible to write down. Gödel had still not found an unprovable statement with actual mathematical content—a statement that a mathematician might actually care about.

And that is where Cantor's Continuum Hypothesis re-enters the picture. Is there a set larger than the integers but smaller than the real numbers? It's a natural question. It seems accessible to our intuition. It's certainly a question a mathematician would care about. And it is an undecidable question. In 1940, Gödel showed that the Continuum Hypothesis *cannot be disproved* from the ZF axioms of set theory. And in 1963, an American mathematician, Paul Cohen, closed the circle by showing that the Continuum Hypothesis also *cannot be proved* from the ZF axioms. In other words, it is logically independent of the other axioms of set theory. You are free to assume it, or you are free to deny it. Neither the Continuum Hypothesis nor its negation will introduce any new paradoxes.

BOTH GÖDEL'S AND COHEN'S arguments proceed by constructing a model of set theory, though I will not explain them in detail. Gödel restricted the universe of possible sets to what he called "constructible sets." Within this model, he showed that the Continuum Hypothesis is true. Therefore, assuming the ZF axioms are consistent, it is impossible to disprove the Continuum Hypothesis. Cohen, on the other hand, worked out a way of extending the universe of possible sets beyond the minimal model allowed by the ZF axioms. Because it involves the creation of new sets, Cohen's argument is more difficult than Gödel's. But the bottom line is that he forced the Continuum Hypothesis to be false … within that model. Again, this does not mean that the "real" Continuum Hypothesis is false, but it does mean that it is impossible to prove the Continuum Hypothesis in all models of ZF set theory.

Cohen's method of "forcing" allowed mathematicians and logicians to discover many other statements that are independent of the ZF axioms. In this sense, it has had a greater impact on mathematics than Gödel's Incompleteness Theorem. The crater is still fresh.

So is the story of the Continuum Hypothesis finished? It's difficult to say. For now, most mathematicians are happy with the ZF axioms, if they even know them. As long as that remains the case, we will have to let the Continuum Hypothesis remain in the ambiguous state where Gödel and Cohen left it. But the ZF axioms may not remain so fashionable forever. Remember that you can never prove an axiom system; it is only a starting point. Someday logicians might devise an axiom system that mathematicians will like better than ZF, and then the Continuum Hypothesis will have to be reconsidered.

After the Second World War, Gödel ended up at the Institute for Advanced Study in Princeton, where he became best friends with Albert Einstein. Einstein once said that on many days the only reason he would come to the Institute was "to have the privilege to walk home with Gödel." However, in later years Gödel's behavior became increasingly odd; for example, he believed in ghosts, and he thought that someone was trying to poison him. He would eat only food that he had personally prepared, and in the end, he practically starved himself to death.

Cohen, thankfully, had none of the struggles with mental health that Cantor and Gödel did. Born in 1934, he grew up in Brooklyn and was a self-taught prodigy. He never finished college but went straight on to graduate school in 1953, at the age of 19. Curiously, he was not trained as a logician, but was attracted to the Continuum Hypothesis because he had a strong intuition that it was false. (That's right, he disagreed with Cantor.)

Sometimes it takes the fresh outlook of a non-expert to break out of an intellectual logjam. Immediately after he discovered the idea of forcing, other logicians seized on it, and Cohen found himself unable to keep up with them. Nevertheless, Cohen had the honor of being first, and his work received high praise from none other than Gödel himself. "You have just achieved the most important progress in set theory since its axiomatization," Gödel wrote to him.

23

theories of chaos
lorenz equations

John von Neumann had a dream. In 1954, speaking at the dedication of the world's new largest computer, Neumann (one of America's first great computer scientists) predicted that computers would one day make it possible to forecast the weather for 30 to 60 days.

More than half a century later, computer power has grown beyond Neumann's wildest dreams. Yet computer weather forecasting has made only incremental improvements. Two-day forecasts are now pretty reliable. Five-day forecasts, considerably less so. And not even the most optimistic meteorologist dreams any more of predicting a storm 60 days in advance.

What went wrong? The answer, in a word, is chaos.

A chaotic system is one that obeys deterministic rules—such as the equations that describe how air circulates in the atmosphere—yet behaves, after a certain amount of time, as if it were random. In theory, if you have perfect information about a chaotic system, you can make perfect forecasts. But the slightest inaccuracy or incompleteness in your data will grow exponentially over time, and eventually render your forecast useless.

Ironically, the one tool that most enabled scientists to grasp the implications of chaos, and thereby to place a permanent limit on the power of computation … was the computer. The computer did something that von

$$\frac{dx}{dt} = -10x + 10y$$

$$\frac{dy}{dt} = 28x - y - xz + 10y$$

$$\frac{dz}{dt} = -\frac{8}{3}z + xy$$

x, *y*, and *z* are meteorological variables in an abstract and highly simplified model of the atmosphere. These equations were historically the first dynamical system in which scientists recognized the possibility of chaos.

Neumann never expected. Instead of merely crunching numbers, it provided humans a new way of looking at the world.

A second irony is that the person who first understood the mathematical concept of chaos was not a mathematician. Edward Lorenz, a meteorologist, discovered the equations above that effectively ended von Neumann's dream.

Lorenz's three simple equations, which describe a highly idealized atmosphere, are shown above. It is worth taking a little bit of time to understand what these equations mean. First, note that they are differential equations: they express the rate of change of three variables (x, y, and z) in terms of their current values. Calculus was built to solve such equations. They are the type of equations Newton and Laplace used to describe the motion of planets, and that *Apollo* engineers used to send rockets to the Moon. For centuries, scientists assumed that the solutions to such equations were manageable and predictable.

Second, note that the equations are nonlinear. They include two terms (xz in the second equation, and xy in the third one) that are not first powers of the variables but products of two variables (which makes them count as second powers). This tiny detail makes all the difference in the world.

Linear equations describe a textbook world, where effects are always proportional to their causes. The whole is always exactly equal to the sum of its parts. A tiny error in measuring one of the variables will remain a tiny error—or at worst, will grow in nice linear fashion—for all time. It is a world where the weather is predictable for weeks, or months, or even forever.

THE REAL WORLD, however, is nonlinear. Feedback loops amplify small causes into big effects. In biology, nonlinearity arises whenever one cell signals another cell to stop working, or start working harder. In chemistry, nonlinearity occurs when one chemical catalyzes a reaction involving another. In aerodynamics, the equations become nonlinear when the air turns into a moving object rather than a passive medium. Most of the interesting phenomena in science, including anything that involves the mediation of one part of a system by another part, are nonlinear.

Below The Lorenz Attractor, a three-dimensional graph which became emblematic of chaos.

In the Lorenz equations, thanks to those two terms xy and xz, the variable x mediates the way that z responds to y, and also mediates the way that

y responds to z. Nevertheless, this nonlinearity seems so mild that nearly every mathematician who has seen these equations has probably thought, "Why, I could solve them!" But they can't.

Now I'll explain what the variables x, y, and z meant to Lorenz when he wrote these equations down in 1963. Lorenz's model describes the convection of air in a long rectangular tube when the bottom is heated. Hot air tends to rise, so eventually a rolling current forms, with hot air rising on one side of the tube and cool air descending on the other. But in Lorenz's model, the convection current eventually starts going too fast. The hot air doesn't have a chance to cool down completely before it gets swept down the other side of the tube. Hot air doesn't like to descend, so this slows down the rolling motion—which eventually stops, and then *switches direction*. These reversals of direction constitute the unpredictable feature of the system.

Lorenz describes the meaning of the variables x, y, and z as follows: x represents the strength of the convective current; y represents the size of the temperature gradient between the ascending and descending streams. The meaning of z is a bit elusive, but it is crucial: "The variable z is proportional to the distortion of the vertical temperature profiles from linearity, a positive value indicating that the strongest gradients occur near the boundaries," Lorenz writes.

When the variables x, y, and z are plotted over time, from almost any starting point, they will eventually coalesce around an intricate, butterfly-shaped structure, shown opposite. The two "wings" of the butterfly correspond to the two directions of rotation of the convection currents. A typical trajectory starts toward the center of one wing and gradually spirals outward. Eventually it goes "too far" out (the convection currents get out of control). At that point the trajectory plunges through the complicated mess between the two wings and emerges on the other side, ready to start its spiraling motion again.

If you start a second trajectory at a slightly different point, it will behave the same way for a little while. Both trajectories may, for instance, make two loops around the left wing, then three loops around the right. But the distance between the trajectories will grow, and then there will come a time where trajectory 1 veers left while trajectory 2 heads right. From then on, the two trajectories will be uncorrelated. It is easy to make the metaphorical leap to weather forecasts. Think of "left" as "sunny" and "right" as "rainy." If trajectory 1 represents the real weather, while trajectory 2 represents a forecast based on slightly different initial conditions, the two may agree for a few days, but eventually the forecast is guaranteed to bear no resemblance to reality.

Before we leave the details of the Lorenz model behind, let me point out one more curious feature: the seemingly arbitrary constants, 10, 28, and $8/3$. These are called "parameters," and they have a strong effect on the shape of the solutions. If you replace the number 28 by 24 (or any number below about 24.8), the chaos disappears; the convection currents are not sufficiently excitable. Starting from any initial state, the convection currents will eventually settle down into a stable state, either rotating left or rotating right. After that, the system becomes 100 percent predictable. Thus, nonlinearity is not a *guarantee* of chaos; it merely opens up the possibility. The tipping point where a regular system becomes chaotic often depends on parameters that we cannot observe.

In one short paper, Lorenz had identified most of the main ingredients of chaos, although he had not named them yet: Sensitive dependence on initial conditions (the "butterfly effect")[‡]; long-term behavior that is controlled by an infinitely complicated (and beautiful!) geometric structure (later called a "strange attractor"); a parameter (or several) that can switch on or switch off the chaos; and nonlinear but completely deterministic dynamics.

The importance of Lorenz's paper was not immediately apparent; it was buried in a specialist journal, read only by meteorologists. However, the same process repeated in other disciplines. Michel Hénon, an astronomer, discovered chaos in the equations governing stellar orbits around a galaxy's center. David Ruelle, a physicist, along with mathematician Floris Takens, discovered strange attractors in turbulent fluid flow. Robert May, a biologist, discovered chaos in the simplest system of all: a one-variable equation that had been used for years to model populations competing for scarce resources.

Each of these pioneers was isolated at first, and they all faced disbelief from other scientists. A colleague of Lorenz, Willem Markus, recalled in James Gleick's bestselling book *Chaos: Making a New Science* what he told Lorenz about his equations: "Ed, we know—we know very well—that fluid convection doesn't do that at all."

This incredulity is perhaps a typical reaction to any paradigm-altering discovery. In the case of chaos there were specific reasons why mathematicians and other scientists had been so blind for so long. When mathematicians teach their students differential equations, they concentrate on the simplest, most understandable cases first. First, they teach them to solve linear

---

[‡] This name came from the title of a paper that Lorenz himself presented in 1972, called "Predictability: Does the Flap of a Butterfly's Wings in Brazil Set Off a Tornado in Texas?"

equations. Next, they might teach them about some simple two-variable systems, and show how the behavior of the solutions near a fixed point can be described by linearizing. No matter what the number of variables, they will always concentrate on equations that can be solved explicitly: $x(t)$ is given by an exact formula involving the time t.

ALL OF THESE simplifying assumptions are perfectly understandable, especially the last one. Solving equations is what mathematicians do ... or *did*, in the years BC (before chaos). And yet these assumptions are collectively a recipe for blindness. Chaos does not occur in linear systems; it does not occur in a continuous-time system with less than three variables;§ and it does not occur in any system where you can write a formula for the solution.

It is as if mathematicians erected a "Danger! Keep out!" sign at all of the gates leading to chaos. Scientists from other disciplines—biologists, physicists, meteorologists—never went past the "Keep out!" signs, and so when they encountered chaos it was something utterly unfamiliar.

A very small number of mathematicians did venture past the warning signs. The first one, universally acknowledged by all chaos theorists, was Henri Poincaré, France's greatest mathematician at the turn of the century. In 1887, he entered an international competition, sponsored by the King of Sweden, to find a solution to the three-body problem, in other words to find explicit formulas for the trajectories of three or more mutually gravitating planets. Isaac Newton had, of course, solved the problem for two bodies, and it had been a bone in the throat of mathematicians ever since that they could not do the same for even the simplest system of three bodies.

Poincaré won the prize even though he did not solve the problem. In fact, he thought he had solved it, but as he was preparing his manuscript for publication (after he had been awarded the prize!) he discovered a mistake. He had assumed that small perturbations in a planet's motion would produce small effects. Analyzing a planet's "return map" more carefully, he realized that was not the case. Thus, he clearly discovered the first hallmark of chaos,

---

§ In discrete-time systems, such as May's equation that describes the change in population of a species from one year to the next, chaos is possible even with only one variable.

the sensitive dependence on initial conditions. Much more obscurely, he sensed the second feature as well, the strange attractor. In the following passage he is describing the trajectories of planets in phase space, but I encourage you to think about the Lorenz attractor as you read it:

"These intersections form a kind of lattice-work, a weave, a chain-link network of infinitely fine mesh; each of the two curves can never cross itself, but it must fold back on itself in a very complicated way so as to recross all the chain-links an infinite number of times ... One will be struck by the complexity of this figure, which I am not even attempting to draw."

He is not describing the Lorenz attractor *per se*, but he might as well be. There, too, we see a phenomenally complex interweaving of curves, a sort of freeway where infinitely many lanes merge and then go off in different directions without colliding. Thus mathematicians had the opportunity to discover chaos in 1893, when Poincaré's book appeared. But they didn't. They were not prepared to look for chaos; the whole point of the prize competition was to look for stable solutions.

Opposite Fractal image of part of the Mandelbrot Set. Fractal geometry is part of the mathematics of chaos, the study of unpredictable dynamical systems.

The other reason mathematicians were blind to chaos was that they had no computers, and were left with the kind of vague description that Poincaré gave, which other mathematicians failed to understand. With a computer, you can't help but see the attractor in all its glory. For scientists like Lorenz and Hénon who were not professional mathematicians, Poincaré's work was inaccessible but the strange attractors were there on their computer printouts, begging to be explained.

BETWEEN 1893 AND 1970, mathematicians assembled some of the ingredients of chaos theory, without managing to bring them all together. Around the same time as Poincaré, Aleksandr Lyapunov in Russia defined the Lyapunov exponent, a measure of the tendency of nearby trajectories to diverge. In the 1930s and 1940s, Mary Cartwright and John Littlewood in England studied the van der Pol equation, an early nonlinear equation used in radio and radar. Littlewood commented on the "whole vista of the dramatic fine structure of solutions." In the 1960s, Steven Smale, an

American mathematician and fervent anti-Vietnam War activist, described very general topological conditions which guaranteed that a dynamical system would approach a complicated limit set. This established that chaos was a generic phenomenon that exists over a wide range of parameter values. Finally, Benoit Mandelbrot in France, for completely different reasons, opened up the world's eyes to the ubiquity of "fractals" in nature. The Lorenz attractor, with a fractional dimension of about 2.07, is a prime example. Mandelbrot put his finger exactly on what was strange about strange attractors: they have fine structure at every scale, so that a magnified version looks just as finely filigreed as an unmagnified version.

IN THE 1970S the grand synthesis occurred. The disparate scientists who had encountered chaos began to find each other and connect with the mathematicians who could explain their discoveries. In 1975 the field acquired its seductive nickname, thanks to a paper by Tien-Yien Li and James Yorke, called "Period Three Implies Chaos." Li and Yorke showed that a one-variable discrete dynamical system, like the one studied by May, must be chaotic if there is even one point that comes back to itself after three time steps. A Ukrainian mathematician, Olexandr Sharkovsky, had proved the same fact eleven years earlier, but no one in the West knew about it because of the lack of communication across the Iron Curtain. Sharkovsky's paper was finally translated in 1995 and the theorem is now known after him, but Li and Yorke had the honor of giving "chaos theory" its name.

In the 1980s and '90s, the trope of "chaos" leaped out of the realm of science and into popular culture. James Gleick brought attention to the field with his best-selling book. In the blockbuster movie *Jurassic Park*, one of the leading characters is a doom-predicting chaos theorist, and chaos is the central metaphor for the disastrous failure of scientists to anticipate the consequences of their actions.

The same period was a high-water mark for the subject, with three major chaos journals launching in 1990 and 1991. Two decades later, the excitement has died down a bit. In fact, I was surprised to read in a recent historical survey that "as a unified site of social convergence, the 'science of chaos' does not exist any more."

I think the obituary is premature. Certainly a number of interesting discoveries have been made in chaos since 1990. Perhaps the most surprising is synchronized chaos. You might expect that two oscillators that both behave in an apparently random way would be impossible to synchronize. Yet it turns out that if you feed just one of the three output variables of one Lorenz oscillator (or any other chaotic system of your choice) into another Lorenz oscillator, both of the other variables in the second oscillator will also lock onto the first oscillator. It is an ingenious way to exploit the deterministic laws that lie hidden underneath the apparent randomness. This may explain how living organisms, such as nerve cells, synchronize their behavior.

Other recent developments include the discovery of limited forms of "quantum chaos." For a long time the equations of quantum mechanics resisted the incursion of chaos because they are linear. It's a good thing they are; we would not want the electrons, protons, and neutrons that make up matter to be unstable. But somewhere in the "quasi-classical limit," the gray zone between the macroscopic world and the quantum world, chaos has to make its appearance, and both mathematicians and physicists have been probing how.

Finally, much work remains to be done in understanding turbulent fluids. Chaos is not the end of the story, but only the beginning. Using Lyapunov exponents, scientists can find the invisible attracting structures and repelling structures that orchestrate fluid motion. They can identify where the flow is chaotic and where it is not. They can map out, for example, the invisible dividing line between water in the Gulf of Mexico that will circulate back into the gulf and water that will escape into the Atlantic. Methods like this could have been used to predict the motion of the BP oil spill in 2010.

So I think it is fair to say that the *concept* of chaos is alive and well, and always will be, now that we have finally learned to see it. The *discipline* of chaos is also alive and well, although it may have outlived the fad stage and will probably end up being seen as an organic part of a more traditionally named discipline, such as "dynamical systems" or "nonlinear differential equations."

taming the tiger
the black–scholes equation

In 2003, Roy Horn—half of the famous performing duo Siegfried & Roy—was bitten on the neck by one of his own tigers. Roy nearly died, and his performing days were over. To fans, it was a shocking reminder that tigers are still tigers. The sheer technical proficiency of Siegfried & Roy had lulled audiences into complacency. As Neil Strauss wrote in *The New York Times*: "Danger was still present, but it was no longer recognized as such."

A similar thing could be said about the Black–Scholes equation. Published in 1973 by Fischer Black and future Nobel Prize winner Myron Scholes (with a big assist from another future Nobelist, Robert Merton), it seemed to take the danger out of investing. It led to an explosive growth in the market for financial products called derivatives—essentially, bets on the direction that certain other assets, such as stocks and bonds,[*] would move. Like the Las Vegas audiences of Siegfried & Roy, Wall Street was dazzled by the technical proficiency with which a new generation of traders, called "quants," wielded the Black–Scholes equation. Danger—or as financial engineers call it, "risk"—seemed to be under control.

---

[*] In the rest of this chapter, for convenience, I will refer to stocks, although stock options are actually far from being the most common kind of derivative.

$$\frac{\partial V}{\partial t} + \frac{1}{2}\sigma^2 S^2 \frac{\partial^2 V}{\partial S^2} + rS\frac{\partial V}{\partial S} = rV$$

V is the market value of a financial derivative called a call option, and S is the value of the underlying asset (e.g., a stock) at maturity. ρ and σ represent the interest rate and the volatility of the stock. The equation gave economists the possibly mistaken impression that risk can be managed according to objective laws that resemble the laws of physics.

But tigers are still tigers, and financial markets are still financial markets. Not once but three times since Fischer and Scholes' breakthrough—in 1987, 1998, and 2007—the market has turned on the people who thought they could control it. No one has been killed, but careers have been ruined and fortunes lost. The credit crisis of 2007–8, in particular, led to the most severe recession in America since the Great Depression of the 1930s.

Some people have tried to blame mathematics, or the quants, for these events. For example, an article in *Wired* in 2009 referred to "the formula that killed Wall Street." But before we can pass judgment, we should first understand what Black, Scholes, and Merton accomplished.

Let's start with a greatly simplified example of a derivative. Suppose that today's price of one share of stock in the Well-A-Day Oil Company is $100. Suppose also that you happen to know that the stock has a 50-50 chance of rising to $101 tomorrow, and a 50-50 chance of dropping to $99. Amazingly, you can use this knowledge to turn a guaranteed profit of 50 cents per share, even though you don't know which way the stock is going to go.

First, you call up your broker and you ask him to give you an option to buy two shares of Well-A-Day stock tomorrow for $100 each. This seems like a reasonable deal, right? After all, there is an equal chance that the price will be above $100 or below $100. (Note: If your broker agrees to this, he is a fool and will go out of business soon, but I'll explain why in a moment.)

Now, armed with your option, you hedge it by borrowing one share of Well-A-Day from your friend Bob and selling it for $100. The next day, there is a 50-50 chance that the stock will go down to $99. In that case, you let the option expire unclaimed, but you can buy one share for $99 and return it to Bob. Your profit is $1, because you bought for $99 and sold for $100. On the other hand, if the stock goes up to $101, then you claim your option and buy two shares from your broker for $100 each. You return one to Bob, and sell the remaining share for $101, the current market price. Once again, your profit is $1, because you bought two shares for $200 and sold them for $201. Thus, no matter what happens, you earn a dollar. It doesn't matter whether the price of the "underlying asset" goes up or down.

THIS HEDGING PRINCIPLE has been rediscovered many times over the years. Prior to 1973, the people who discovered it usually thought they had discovered the secret to getting rich. In a typical book called *Beat the Market!* from 1967, economist Sheen Kassouf writes about his epiphany, "I realized that an investment could be made that seemed to insure tremendous profit whether the common rose dramatically or became worthless. I would win whether the stock went up or down! It looked too good to be true."

Of course, like all get-rich-quick schemes, it *is* too good to be true, for two reasons. First, your broker will get wise to this game pretty fast. In fact, the hedging principle shows that the option to buy one share of Well-a-Day for $100 tomorrow is actually worth 50 cents today. The option itself has value. However, in the early days of option trading, brokers and investors did not have a very good idea of how to price real-world options (as opposed to this made-up example). When Kassouf and his co-author Edward Thorp wrote their book, they could sift through the published prices of stock warrants (the most common type of option available then) and find mispriced warrants. A savvy investor *could* beat the market.

Secondly, even if we can find a broker who will sell us the option for the wrong price—say, 49 cents instead of 50 cents—we will realize our profit (which is now down to a cent per share!) only if we are absolutely right

Above Symbols of the stock exchange: The bull and bear statues outside the stock exchange in Frankfurt, Germany.

about our prediction that the stock will either go up to $101 or down to $99. Unlike normal investors, who base their investments on the *direction* that the stock is going to move, we don't care about the direction. But we do care about the *volatility*—the size of the jump. Given the narrow margin for error, the hedging principle will only work if our estimate of the stock's volatility is right on the money.

However, this critical point was hidden by the technical virtuosity of the Black–Scholes formula. Suppose that we want to find the fair market value, V, of a call option—an option to buy Well-A-Day stock at price K (the "strike price") at time T (the "expiration date"). We don't want to cheat anyone; we just want to know how much this option is worth.

Clearly, the option's value depends on the current stock price, S. The higher the current price, the more likely it is that the stock price will still be above K on the expiration date. Also, the value depends on the time remaining until expiration, $T - t$. Even if the option is "out of the money" today, more time gives the stock more of a chance to rise to the strike price. Finally, *on* the expiration date, the value of the option is either 0 (if the stock price is below K) or else it is $S - K$ (because if the stock price is above K, we will exercise the option and buy it for K dollars, then go to the open market and sell it for S dollars). Thus at time T, the option's value looks like the solid line in the graph below. At earlier times t, before the expiration date, the option's value should be a little bit higher, like the dotted line in the figure. But how much higher?

Black, Scholes, and Merton proved that there *is* a fair market value $V(S, t)$. This is pretty surprising, because you might think that an investor who is bullish on Well-A-Day would value the option differently from someone who is bearish. Their main idea was to use dynamic hedging to eliminate risk. This is just a more elaborate version of the hedging principle, in which the investor has to adjust his or her portfolio constantly, in accordance with the current price S and time t. The effect is the same: dynamic hedging makes it irrelevant whether we personally believe the stock will go up or down.

Thus the bulls and the bears can agree on the value of the option—provided that they agree on a model for the volatility. And that was the second ingenious stroke of Black and Scholes. They proposed a model of stock prices that was so "intuitively obvious" that hardly anyone could disagree with it. Changes in stock prices, they said, have two components: an upward or downward drift plus a random jiggle. It's only the size of the jiggle, the volatility, which matters for the option price, and this volatility is measured by a number sigma (σ). A well-known function, called the normal distribution or the bell-shaped curve (see opposite), gives the likelihood of any particular size of fluctuation. For example, the likelihood of a "one-sigma" (or more) increase in the price is about 15.8 percent, and the likelihood of a "two-sigma" increase is about 2.2 percent.

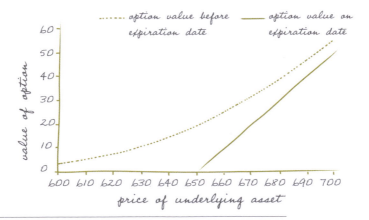

Above On the expiration date, the value of an option to buy stock at $650 looks like the solid line. Because of dynamic hedging strategies, the value of the option before the expiration date is always higher (dotted line).

Why was the Black–Scholes proposal so seductive? Perhaps because the bell-shaped curve is so familiar. Any stochastic process that amounts to infinitely many independent flips of a coin (even a biased coin) will produce a normal distribution. If you've ever watched a stock ticker, you have probably seen an endless stream of pluses and minuses scroll past, indicating upticks and downticks of the stock. It really does look like an infinite (or nearly infinite) succession of coin flips.

FINALLY, BLACK AND SCHOLES had one more stroke of genius. Unlike previous investors who saw the hedging strategy as a way to beat the market, they insisted that there is no way to beat the market. If you hedge your portfolio in a way that eliminates risk, you should get exactly the same rate of return as someone who invests in the most risk-free investment—a 30-year US Treasury bond. This argument closed the loop and gave them the Black–Scholes equation:

$$\frac{\partial V}{\partial t} + \frac{1}{2}\sigma^2 S^2 \frac{\partial^2 V}{\partial S^2} + rS\frac{\partial V}{\partial S} = rV$$

In this equation, the left-hand side represents the return on your investment if you buy the option and hedge it dynamically according to Black and Scholes' prescription. The right-hand side represents the return if you

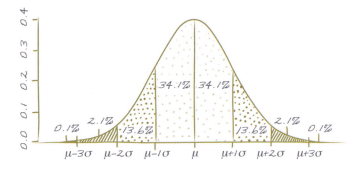

Above The standard deviation, denoted σ, is a measure of the amount of spread in a classic bell-shaped curve.

simply put it in the bank (with r representing the interest rate). According to Black and Scholes, in an efficient market the two returns are equal.

Like Maxwell's equations and the heat equation, the Black-Scholes equation is a partial differential equation, a type that physicists and mathematicians are very aware of. The same sort of equation describes the diffusion of molecules in a gas, because their motion likewise consists of innumerable tiny jiggles. For the simplest call options, Black and Scholes deduced an exact solution for the value V. However, the Black–Scholes equation also applies to all sorts of other, more "exotic" options, such as options based on more than one underlying stock or options based on mortgage defaults.

Remarkably, one mathematician had anticipated Black and Scholes by more than 70 years. In 1900, a student of Poincaré named Louis Bachelier had studied an almost identical model of option prices—including the same idea of random fluctuations—and derived an almost identical equation. But Bachelier lived at the wrong time. The discipline of mathematical economics did not yet exist. Pure mathematicians were very interested in his idea of a process consisting of infinitely many small jiggles (now called "Brownian motion"). However, they were not the least bit interested in Bachelier's motivating example. One colleague, Paul Lévy, wrote a disparaging comment in his personal notebook: "Too much on finance!"

But by 1973, the world was ready. That year, the world's first options exchange opened in Chicago. Over the next two decades, the Black–Scholes formula changed Wall Street. First of all, it created jobs for a whole new kind of trader—a "quant," usually someone with a mathematics or physics background who understood differential equations. But perhaps more importantly for society, Black–Scholes created an aura of invincibility around mathematical finance. Options, once seen as a somewhat disreputable investment (a high-risk wager on the stock market), now seemed to be the exact opposite. They were an essential tool for controlling or eliminating risk. For Black, Scholes, and Merton, it was an article of faith that the world was gradually moving toward an ideal state where you truly couldn't beat the market, and all options would be rationally priced according to the mathematical models. From this point of view, quants weren't just making money; they were helping to make the market more efficient.

Just like von Neumann's dream of perfect weather forecasting, the dream of perfect markets never came true. The first crack appeared in 1987, on Black Monday, when the Dow Jones Industrial average dropped by more than 22 percent, by far its biggest one-day percentage loss ever. It was widely blamed on programmed trading—the kind of automatic selling of stock that dynamic hedging requires.

The second crack, in 1998, was much more personally embarrassing to the theory's founders. By this times, Scholes and Merton were both partners in a hedge fund called Long-Term Capital Management (LTCM). Although they were not involved in day-to-day operations, the two Nobel Prize winners lent a huge amount of prestige to the fund, which was like a laboratory experiment in risk-neutral investment.

Between 1994 and 1998, LTCM quadrupled its investors' money and seemed to live up to the promise of guaranteed returns. Then, over a span of less than two months, it all came crashing down. A cascade of events, starting with the Russian government defaulting on its debts, pushed volatilities to stratospheric levels, far beyond where the models said they should have gone.

LTCM hemorrhaged money. Finally, with the company on the brink of bankruptcy, the Federal Reserve Bank organized a bailout by a consortium of fourteen leading private banks—a move the banks agreed to, reluctantly, because they feared that LTCM was "too large to fail." If it went down, they could fall like dominos.

FINALLY AN EVEN GREATER CRISIS rocked the financial world in 2007. This time it was a collapse in the market for credit derivatives— one of the more exotic kinds of derivatives mentioned earlier. This time, the Black–Scholes formula was more peripherally involved. For several years, banks had been offering "subprime" loans to home buyers who would not have qualified in the past. The banks were not motivated by altruism; they believed that they could control the risk of default by lumping many mortgages into one security called a collateralized debt obligation (CDO). The quants had developed a convenient formula based on the normal distribution (called the "Gaussian copula") that gave a fair market value for the CDO's … provided that the correlation between loan defaults was zero (or at least not too big). In other words, if a homeowner in Miami forecloses, it shouldn't affect a homeowner in Las Vegas.

But in 2007 the housing bubble burst, and a wave of foreclosures swept the entire nation. Suddenly Miami affected Las Vegas and vice versa. During a panic, all the correlations go to one. Banks across the country did not have enough capital to cover their bets, and one after another they started failing: Bear Stearns, Washington Mutual, Lehman Brothers. Once again the government had to intervene, only in a much bigger way than before. The Secretary of the Treasury announced a $700 billion bailout, or "Troubled Asset Relief Program." In an unprecedented move, the government actually acquired one of the "too-big-to-fail" companies—AIG, the world's largest insurance firm. Unlike in 1998, the private sector did not have enough healthy banks left to do the job.

The common denominator in all of these debacles was the failure of mathematical models to anticipate the volatility of the markets. The normal distribution is based on the assumption of innumerable small actors making innumerable random choices, all independently of each other. But when one actor gets too big (like LTCM) or when the actors stop behaving independently (as in the panic situations of 1987 and 2007–8), the model does not apply. In fact, a small group of dissident economists has argued for years that models like Black–Scholes should never be used because they underestimate the probability of extreme events.

However, the Black–Scholes equation and the philosophy behind it are by now too ingrained to just throw them out. Instead, economists are trying to come up with refinements that better reflect how real markets behave. For example, in the "jump diffusion" model, stock prices have three components: long-term drift, short-term stochastic jiggles, and intermittent jumps due to lurches of the stock market as a whole. This certainly seems to agree better in a qualitative way with reality. Black–Scholes does work very nicely most of the time, aside from the random infrequent occasions when it doesn't work at all. Another approach is to view the volatility σ as being given by a stochastic process itself, or by an empirical function of S and t. Unfortunately, all of these ideas have a kludgy feel to them. They seem contrived to preserve the outward appearance of the Black–Scholes formula without maintaining its internal consistency.

In the end, the question remains: Can mathematics tame the tiger that is the future? Or will it always break out of its cage just when we least expect it? I cannot pretend to answer this question; it is something for twenty-first century economists, financial engineers, and mathematicians to work on. Until they succeed, or else prove a new Impossibility Theorem, the lesson from 1987, 1998, and 2007 is definitely "buyer beware."

conclusion
what of the future?

The end of this book invites a question: What comes next for equations?

First, the good news. The enterprise of mathematics and science worldwide seems still to be in very healthy shape. There seems to be an upward trend in the sheer quantity of important formulas, which mirrors the growth of mathematics and science in our society. When I began working on this book, in 2008, I searched an authoritative website, Wolfram MathWorld for three terms: "equation," "formula," and "identity." The search engine dutifully reported 1947 equations, 1253 formulas, and 992 identities. Three years later, the same search yields 2032 equations, 1307 formulas, and 1026 identities.

Another positive development is the Internet, which has made the sharing of scientific ideas so much easier. Remember how in the "bad old days" of the 1500s and 1600s, mathematical progress was repeatedly slowed by the reluctance of researchers to share their work. And even in the 1900s, politics prevented some of the work of Soviet mathematicians from becoming widely known in the West. Now, thanks to such websites as the e-print archive, as well as mathematical forums and blogs, the barriers to communication are lower than they ever have been.

On the other hand, not all of the signs are positive. Indeed, there are some reasons to think that the twentieth century may have been a high-water mark for interesting, consequential, and beautiful equations.

First, although we have added 173 equations, formulas, and identities in the last three years, there is no guarantee that they meet the criteria I listed in the Introduction. Are they surprising, concise, consequential, and universal? Does the twenty-first century really have such monumental surprises in store as quantum physics, chaos, or Gödel's Incompleteness Theorem? It's impossible to predict, but to me those look like oceans that can only be crossed once. Geographers eventually ran out of new continents on Earth to discover, and it seems possible that mathematicians will face the same problem.

Also, changes are afoot in the way mathematics is conducted. The computer has brought humans a new way of knowing. Forecasting the climate or mapping the human genome involve the collection of staggering amounts of data—amounts too great for the human mind to comprehend. Scientists have to devise new ways to sift through the data and identify what is important. The most important patterns may not be expressible any more in the form of an equation. Perhaps they will be encoded into an artificial intelligence and not even understandable by human brains at all.

Let me give a specific example where this change has already happened. Over the last 25 years, chess players' ways of knowing has been dramatically changed by the computer. For example, there are positions that a computer can solve but a human cannot. Perhaps the most technical endgame that humans can master is checkmate with a king, bishop, and knight against a lone king. The procedure is tricky, but it can be broken down into stage, and it takes at most 33 moves (according to the computer) if both sides play their best. However, computers have discovered other endgames, such as king, rook, and bishop against king and two knights, where it takes as long as 223 moves for the stronger side to win, assuming both players play perfectly. And the moves are completely incomprehensible to humans. When you compare the position after move 80 to the position after move 50, it is impossible to explain why the stronger side is 30 moves closer to victory.

The point of this example is that some kinds of knowledge are too complicated for the human mind to grasp. They are not necessarily deep, just intricate. Equations have evolved as a powerful tool that enables us to grasp some ideas that cannot (or can only with great difficulty) be put into words. But the truths lurking in the databases of the twenty-first century may not be understandable even with equations. They may be the scientific equivalent of the 223-move checkmate.

Finally, I expect some changes in the ways mathematics is used. Historically, mathematics has been tied closely to physics, but in the next century we are likely to see more applications to other disciplines, such as biology or social sciences. The idea of using math to cure cancer is tremendously exciting.

But there is a catch. In order to say anything about the universe with mathematics, we have to construct a mathematical model. And models are always imperfect. First, they always oversimplify reality in some way; and second, every mathematical model begins with assumptions. Sometimes they seem so obvious or so well established by experiment that we forget they are assumptions. We fall in love with our models, and then a major trauma ensues when we have to modify or discard them. Think again of non-Euclidean geometry or the discovery of chaos in deterministic dynamical systems.

For reasons that are not entirely understood, mathematical models historically have worked remarkably well in physics. However, in biology that is not likely to be the case. As a rule, any mathematical model that describes biological processes with any degree of fidelity will tend to fail the conciseness test. It will have many equations and it will be difficult to grasp the reasons for even the most fundamental behaviors. We start entering the domain of the 223-move checkmate. For example, mathematical biologists have developed computer simulations of the heart, which can reproduce such conditions as ventricular tachycardia and fibrillation. Nevertheless, there is not yet any agreement on why the most basic treatment, a defibrillator, actually works.

In sum, I have no doubt that a sequel to this book written a hundred years from now will include six pretty wonderful equations from the twenty-first century. Whether they will be quite as wonderful as Einstein's equations, or Dirac's equation, or chaos, I'm not so sure. We will probably have major breakthroughs in mathematical biology that completely fail the conciseness test. There will be many discoveries analogous to the 223-move checkmate, which cannot be expressed either in words or equations but have to be encoded into an artificial intelligence. The whole idea of an equation might begin to look a little bit quaint.

However, let's not forget that mathematics has an extraordinarily long tradition. Certain things do not change rapidly. One hundred years from now, I predict that there will still be few things quite as satisfying as filling in both sides of an equals sign.

$$? = ?$$

acknowledgments

I will always think of this book in the way that one thinks of a beloved pet that shows up on the doorstep one day, bedraggled and wagging its tail, not certain what it wants but absolutely certain that you are the person that can provide it. Elwin Street Productions had conceived the idea for a book about the history of mathematical equations and they started looking around for a writer. That's when the synopsis landed on my doorstep.

The proposed contents actually made me mad; I had a completely different view of what should be in the book. It also took a while to get used to writing a book that began as somebody else's idea. But in truth, I had been waiting a long time to write a math book, and I knew that I could fix this one up.

So I would first like to thank the people at Elwin Street who tolerated my wholesale changes to their concept, never lost faith in it, and eventually placed the book with the best co-publishers I could ever have asked for, Princeton University Press and NewSouth Publishing.

I would also like to thank, in no particular order:

John Wilkes, the founder of the Science Communication Program at the University of California at Santa Cruz, who has been a guiding light to so many scientists who wanted to become writers.

Peter Radetsky, one of my teachers at UCSC, who told me, "I think you're going to have a great adventure."

Peter Steinhart, who taught essay writing at UCSC, and Rosalind Reid, my first editor at American Scientist, for convincing me that the first person singular pronoun has a place (and a very important one) in science writing.

Martin Gardner, the first popular math writer I ever read, who made it look so effortless.

George Pólya, whose explanation of Euler's Basel Formula (which I read in college) was like Lake Tahoe: so transparent and yet so deep.

Nisaba, the Sumerian goddess of writing (and indirectly, of mathematics), who got the whole thing started.

And Kay, my wife, who encouraged me to follow my dream of writing and then followed the same dream herself. I have learned from her that writing is much more than putting words on paper.

bibliography

Introduction: The Abacist Versus the Algorist

Feynman, R. P. and R. Leighton. *Surely You're Joking, Mr. Feynman!*, New York: W. W. Norton, 1997.

Part One: Equations of Antiquity

Boyer, C. *A History of Mathematics*, 2nd ed. New York: Wiley, 1991.

Burnyeat, M. F. "Other lives," *London Review of Books*, 22 February 2007.

Burton, D. *The History of Mathematics: An Introduction*, Boston: Allyn and Bacon, 1985.

Cipra, B. "Digits of Pi," in *What's Happening in the Mathematical Sciences*, Vol. 6. Providence, RI: American Mathematical Society, 2006.

Dauben, J. "Ancient Chinese mathematics: The *Jiu Zhang Suan Shu* vs. Euclid's *Elements*," *International Journal of Engineering Science* **36** (1998), Nos. 12–14, pp. 1339–1359.

Heath. T. L., ed. *The Works of Archimedes*, Cambridge, UK: Cambridge University Press, 1897.

Huffman, C. "Pythagoreanism," in Stanford Encyclopedia of Philosophy, accessed online at http://plato.stanford.edu/entries/pythagoreanism.

Iamblichus, "Life of Pythagoras," in K. S. Guthrie, ed. *The Complete Pythagoras* (1921), edited for the Internet by P. Rousell and accessed online at http://www.completepythagoras.net.

Kaplan, R. and E. Kaplan. *Hidden Harmonies: The Lives and Times of the Pythagorean Theorem*, New York: Bloomsbury Press, 2011.

Katz, V. J., ed. *The Mathematics of Egypt, Mesopotamia, China, India, and Islam: A Sourcebook*, Princteon, NJ: Princeton University Press, 2007.

Lee, H. D. P., ed. *Zeno of Elea: A Text, with Translation and Notes*, Cambridge, UK: Cambridge University Press, 1936.

Stillwell, J. *Mathematics and its History*, New York: Springer-Verlag, 1989.

Part Two: Equations in an Age of Exploration

Beer, A. and P. Beer, eds. "Kepler: Four Hundred Years," *Vistas in Astronomy*, Vol. 18. Oxford: Pergamon, 1975.

Bellhouse, D. "Decoding Cardano's *Liber de Ludo Aleae*," *Historia Mathematica* **32** (2005), pp. 180–202.

Bradley, R. E., L. A. D'Antonio, and C. E. Sandifer, eds. *Euler at 300: An Appreciation*, Washington, DC: Mathematical Association of America, 2007.

Chambers, R. "Sir Isaac Newton and the Apple," in *The Book of Days: A Miscellany of Popular Antiquities*, pp. 757–8. London: W. & R. Chambers, 1832.

Cipra, B. "Fermat's Theorem—at Last!" in *What's Happening in the Mathematical Sciences*, Vol. 3. Providence, RI: American Mathematical Society, 1996.

Dunham, William. Journey Through Genius. New York: Penguin Books, 1991.

Ekert, A. "Complex and unpredictable Cardano," *International Journal of Theoretical Physics* **47** (2008): pp. 2101–2119.

Grattan-Guinness, I., ed. *Landmark Writings in Western Mathematics*, Amsterdam: Elsevier, 2005.

Gray, J. "A short life of Euler," *BSHM Bulletin* **23** (2008), pp. 1–12.

Montagu, A. "Newton's Principia: Heroes, Legends, and Myths," *The New York Times*,

18 April 1987.

Ouellette, J. *The Calculus Diaries*, London: Penguin Books, 2010.

Rickey, V. F. "Isaac Newton: Man, Myth, and Mathematics," *The College Mathematics Journal* **18** (1987), pp. 362–389.

Stedall, J. *Mathematics Emerging: A Sourcebook 1540–1900*, Oxford, UK: Oxford University Press, 2008.

Weil, A. *Number Theory: An Approach through History from Hammurapi to Legendre*, Boston: Birkhauser, 1984.

Wells, D. Are These the Most Beautiful? Mathematical Intelligencer 12 (1990), No. 3, 37-41.

Westfall, R. *Never at Rest: A Biography of Isaac Newton*, Cambridge, UK: Cambridge University Press, 1983.

Wikipedia. "Principia Mathematica," article accessed online at http://en.wikipedia.org/wiki/Principia_Mathematica.

Part Three: Equations in a Promethean Age

Crease, R. P. "The Greatest Equations Ever," *Physics World*, October 6, 2004. Accessed online at http://physicsworld.com/cws/article/print/20407.

Greenberg, M. J. *Euclidean and Non-Euclidean Geometries: Development and History*, 4th Ed. New York: W. H. Freeman, 2007.

Hankins, T. L. *Sir William Rowan Hamilton*, Baltimore: Johns Hopkins University Press, 1980.

Harman, P. M. *The Natural Philosophy of James Clerk Maxwell*, Cambridge, UK: Cambridge University Press, 1998.

Kaharl, V. A. "Sounding Out the Ocean's Secrets," in *Beyond Discovery: The Path from Research to Human Benefits*. Washington, DC: National Academy of Sciences, March 1999. Accessed online at http://www.beyonddiscovery.org.

Keston, D. A. "Joseph Fourier—Politician and Scientist," accessed online at http://www.todayinsci.com/F/Fourier_JBJ/FourierPoliticianScientistBio.htm.

Lambek, J. "If Hamilton Had Prevailed: Quaternions in Physics," *Mathematical Intelligencer* **17** (1995), No. 4, pp. 7–15.

Laugwitz, D. *Bernhard Riemann, 1826–1866: Turning Points in the Conception of Mathematics*, Boston: Birkhauser, 1999.

Rockmore, D. *Stalking the Riemann Hypothesis: The Quest to Find the Hidden Law of Prime Numbers*, New York: Pantheon, 2005.

Rothman, T. "Genius and Biographers: The Fictionalization of Évariste Galois," *American Mathematical Monthly* **89** (1982), No. 2, pp. 84–106.

Stubhuag, A. *Niels Henrik Abel and his Times*, Berlin, New York: Springer, 2000.

Tolstoy, I. *James Clerk Maxwell: A Biography*, Edinburgh: Canongate, 1981.

Part Four: Equations in Our Own Time

Albers, D. J. and G. L. Alexanderson, eds. *Mathematical People: Profiles and Interviews*, Boston: Birkhauser, 1985.

Albers, D. J., G. L. Alexanderson, and C. Reid, eds. *More Mathematical People: Contemporary Conversations*, Boston: Harcourt Brace Jovanovich, 1990.

Anon. "Lights All Askew in the Heavens," *The New York Times*, 10 November 1919.

Aubin, D. and A. D. Dalmedico. "Writing the History of Dynamical Systems and Chaos: Longue Durée and Revolution, Disciplines, and Culture," *Historia Mathematica* **29** (2002), pp. 273–339.

Brown, K. "Reflections on Relativity," accessed online at http://www.mathpages.com/

rr/rrtoc.htm.

Casti, J. L. and DePauli, *W. Gödel: A Life of Logic,* Cambridge, MA: Perseus, 2000.

Courtault, J.-M., Y. Kabanov, B. Bru, P. Crépel, I Lebon, A. le Marchand. "Louis Bachelier on the *Centenary of Theorie de la Speculation,*" *Math. Finance* **10** (2000), No. 3, pp. 341–353.

Dirac, P. A. M. "The Evolution of the Physicist's Picture of Nature," *Scientific American,* May 1963. Accessed online at http://bit.ly/dirac1963.

Duffie, D. Black, "Merton and Scholes—Their Central Contributions to Economics," in *Legacy of Fischer Black,* B. Lehmann, ed. Cary, NC: Oxford University Press, 2004, pp. 286–298.

Einstein, A. "Does the Inertia of a Body Depend on its Energy Content?" *Annalen der Physic* **18** (1905), p. 639. Trans. by W. Perrett and G. B. Jeffery (1923), accessed online at http://www.fourmilab.ch/etexts/einstein/E_mc2/www.

Friederich, B. and D. Herschbach. "Stern and Gerlach: How a Bad Cigar Helped Reorient Atomic Physics," *Physics Today,* December 2003: pp. 53–59.

Giulini, D. and N. Straumann. "Einstein's Impact on the Physics of the Twentieth Century," *Stud. Hist. Phil. Modern Physics* **37** (2006), pp. 115–173.

Goddard, P., ed. *Paul Dirac: The Man and His Work*, Cambridge, UK: Cambridge University Press, 1998.

Isaacson, W. *Einstein: His Life and Universe*, New York: Simon and Schuster, 2007.

Janssen, M. "Of pots and holes: Einstein's bumpy road to general relativity," *Ann. Phys. (Leipzig)* **14** (2005), Supp. pp. 58–85.

Kassouf, S. T. and Thorp, E. O. *Beat the Market: A Scientific Stock Market System*, New York: Random House, 1967.

Lowenstein, R. *When Genius Failed: The Rise and Fall of Long-Term Capital Management*, New York: Random House, 2000.

Ott, E. "Edward N. Lorenz, 1917–2008," *Nature* **453**, 15 May 2008, p. 300.

Peitgen, H.-O., H. Jergens, D. Saupe. *Chaos and Fractals: New Frontiers of Science*, New York: Springer, 2004.

Pogge, R. W. "Real-World Relativity: The GPS Navigation System," accessed online at http://www.astronomy.ohio-state.edu/~pogge/Ast162/Unit5/gps.html.

Renn, J. and D. Hoffmann. "1905—A Miraculous Year," *Jour. Physics B: Atomic, Molecular, and Optical Physics* **38** (2005), S437–S448.

Schaeffer, S. M. "Robert Merton, Myron Scholes, and the Development of Derivative Pricing," *Scand. Jour. Of Economics* **100** (1998), No. 2, pp. 425–445.

Strogatz, S. *Sync: The Emerging Science of Spontaneous Order*, New York: Hyperion, 2003.

Wikipedia. "The Photoelectric Effect," accessed online at http://en.wikipedia.org/wiki/Photoelectric_effect.

Wilczek, F. "The Dirac Equation," in *Proceedings of the Dirac Centennial Symposium*, Singapore: World Scientific Publishing, 2003.

Wilmott, P. *Frequently Asked Questions in Quantitative Finance*, Chichester, UK: Wiley, 2007.

Woodin, W. H. "The Continuum Hypothesis, Part I," *Notices of the American Mathematical Society*, June/July 2001, pp. 567–576.

Wu, H. "Historical development of the Gauss-Bonnet theorem," *Science in China Series A: Mathematics* **51** (2008), No. 3, pp. 1–8.

Wu, H. "Shiing-Shen Chern: 1911–2004," *Bull. Amer. Math. Soc.* **46** (2009), pp. 327–338.

index